# THE SERGEANT
## A Biological Missile

By
Reid Kirby

Published by

Preneur.Media LLC
7635 W. 148th Street
Suite 263
Apple Valley, Minnesota 55124
United States

*http://www.reidkirby.com*



This book is printed on acid-free paper.

Printed in the United States of America

(Second Printing)

Kirby, Reid.

The Sergeant: A Biological Missile.

Illustrated.

1. Textbook, 2. Biological Warfare. 3. Histroy.

***ISBN: 978-0-9677264-5-8***

*For Brian*

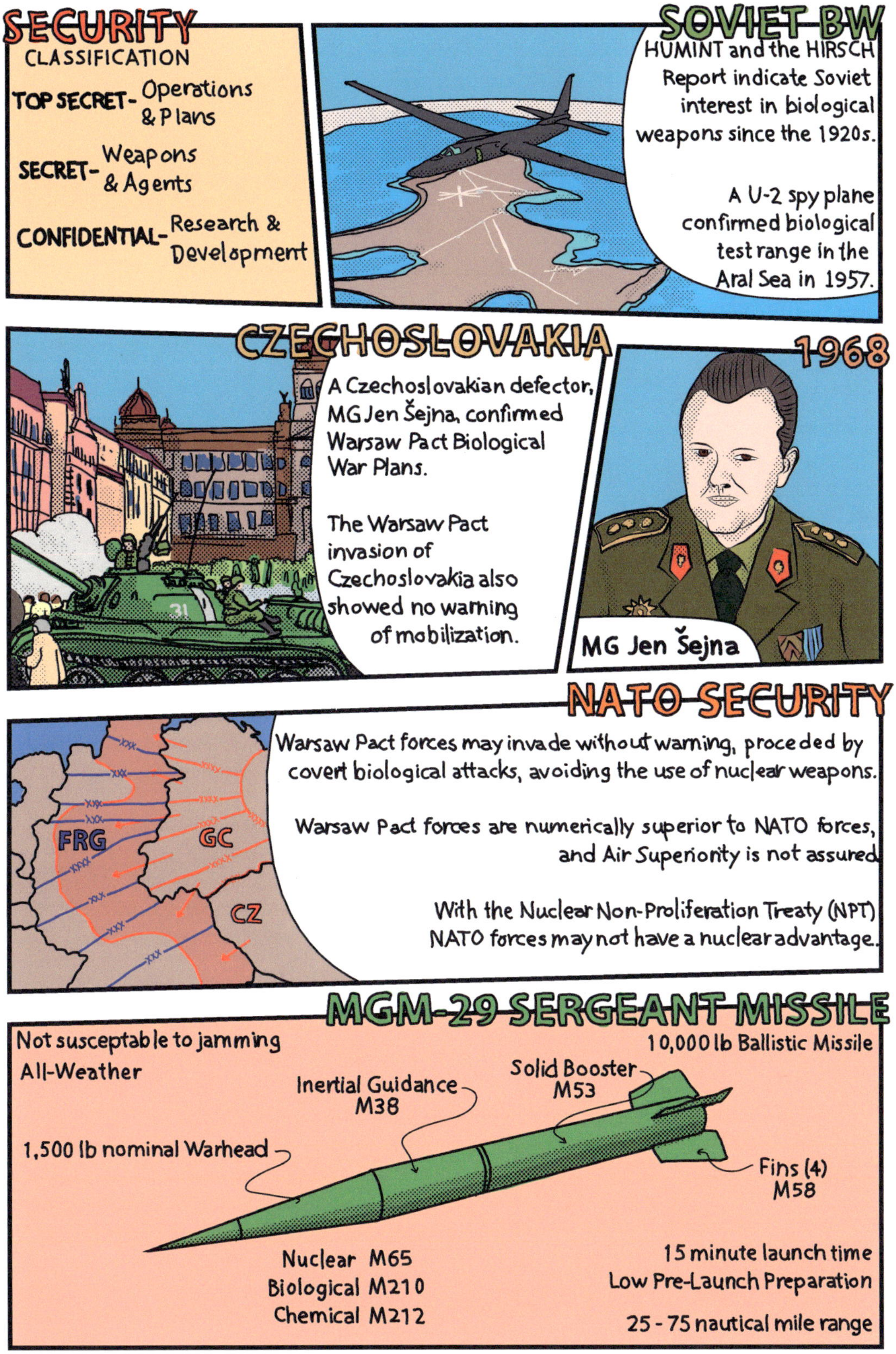
SECURITY
CLASSIFICATION
TOP SECRET - Operations & Plans
SECRET - Weapons & Agents
CONFIDENTIAL - Research & Development
SOVIET BW
HUMINT and the HIRSCH Report indicate Soviet interest in biological weapons since the 1920s.
A U-2 spy plane confirmed biological test range in the Aral Sea in 1957.
CZECHOSLOVAKIA
A Czechoslovakian defector, MG Jen Šejna, confirmed Warsaw Pact Biological War Plans.
The Warsaw Pact invasion of Czechoslovakia also showed no warning of mobilization.
31
1968
MG Jen Šejna
NATO SECURITY
FRG
GC
CZ
Warsaw Pact forces may invade without warning, proceded by covert biological attacks, avoiding the use of nuclear weapons.
Warsaw Pact forces are numerically superior to NATO forces, and Air Superiority is not assured.
With the Nuclear Non-Proliferation Treaty (NPT) NATO forces may not have a nuclear advantage.
MGM-29 SERGEANT MISSILE
Not susceptable to jamming
All-Weather
10,000 lb Ballistic Missile
Solid Booster M53
Inertial Guidance M38
1,500 lb nominal Warhead
Fins (4) M58
Nuclear M65
Biological M210
Chemical M212
15 minute launch time
Low Pre-Launch Preparation
25 - 75 nautical mile range

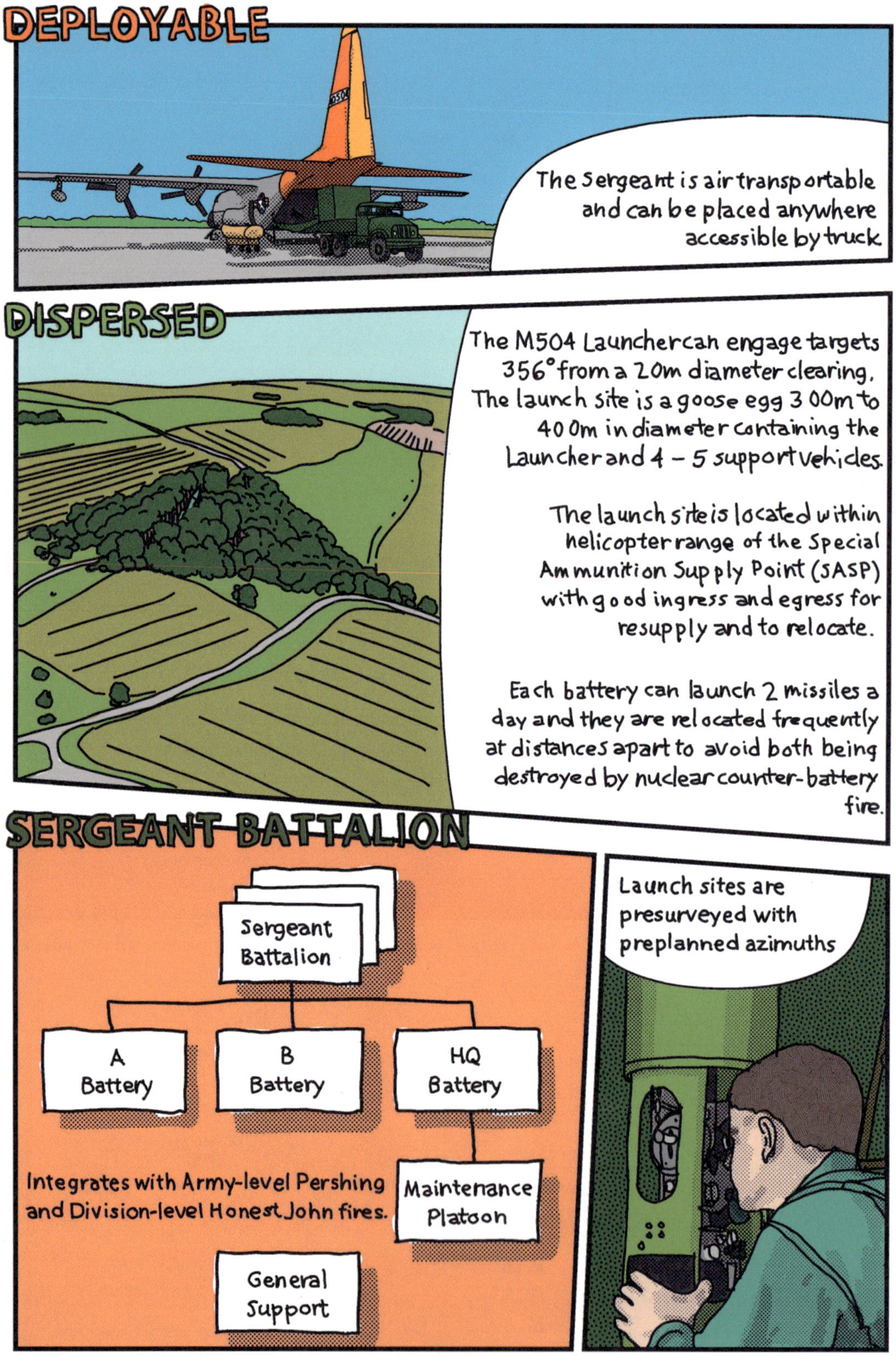
DEPLOYABLE
The Sergeant is air transportable and can be placed anywhere accessible by truck
DISPERSED
The M504 Launcher can engage targets 356° from a 20m diameter clearing. The launch site is a goose egg 300m to 400m in diameter containing the Launcher and 4 – 5 support vehicles.
The launch site is located within helicopter range of the Special Ammunition Supply Point (SASP) with good ingress and egress for resupply and to relocate.
Each battery can launch 2 missiles a day and they are relocated frequently at distances apart to avoid both being destroyed by nuclear counter-battery fire.
SERGEANT BATTALION
Sergeant Battalion
A Battery
B Battery
HQ Battery
Maintenance Platoon
Integrates with Army-level Pershing and Division-level Honest John fires.
General Support
Launch sites are presurveyed with preplanned azimuths

BIOLOGICAL
M210 Warhead
150 kg TT
30% Delayed Casualties
6.5 km Radius Target
131 square kilometers
NUCLEAR
M63 Warhead
200 kt TNT eq. (W52)
20% Prompt Casualties
10% Delayed Casualties
20% Mod. Dest. Tanks
2.8 km Radius Target
25 square kilometers
CHEMICAL
M212 Warhead
195 kg GB
30% Casualties
0.75 km Radius
1.8 sq. kilometers
US ARMY
The Sergeant Is A Corps Ballistic Missile

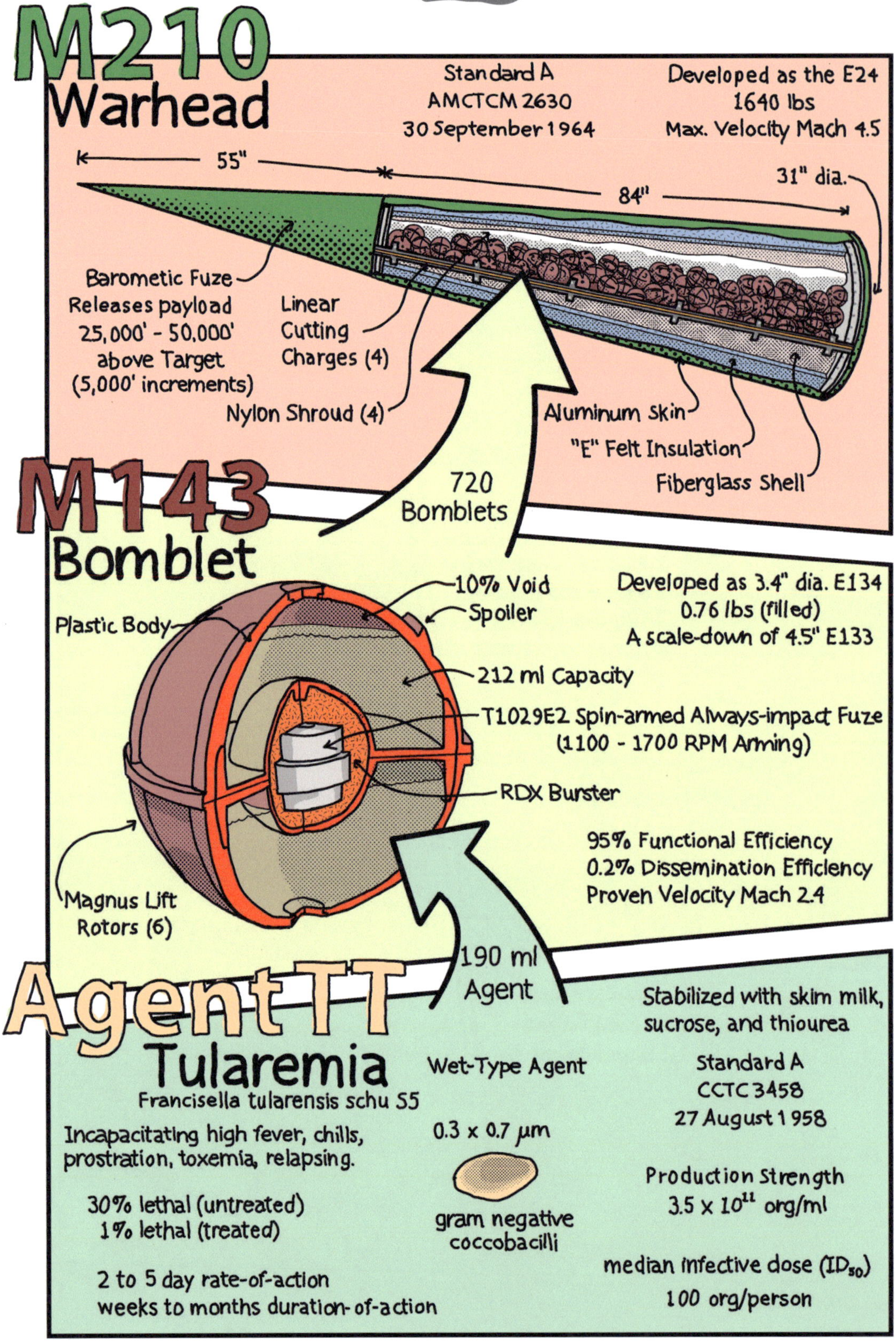
M210
Warhead
Standard A
AMCTCM 2630
30 September 1964
Developed as the E24
1640 lbs
Max. Velocity Mach 4.5
55"
84"
31" dia.
Barometic Fuze
Releases payload
25,000' - 50,000'
above Target
(5,000' increments)
Linear
Cutting
Charges (4)
Nylon Shroud (4)
Aluminum Skin
"E" Felt Insulation
Fiberglass Shell
720
Bomblets
M143
Bomblet
10% Void
Spoiler
Plastic Body
Developed as 3.4" dia. E134
0.76 lbs (filled)
A scale-down of 4.5" E133
212 ml Capacity
T1029E2 Spin-armed Always-impact Fuze
(1100 - 1700 RPM Arming)
RDX Burster
95% Functional Efficiency
0.2% Dissemination Efficiency
Proven Velocity Mach 2.4
Magnus Lift
Rotors (6)
190 ml
Agent
Agent TT
Tularemia
Francisella tularensis schu S5
Wet-Type Agent
Stabilized with skim milk,
sucrose, and thiourea
Standard A
CCTC 3458
27 August 1958
Incapacitating high fever, chills,
prostration, toxemia, relapsing.
0.3 x 0.7 µm
30% lethal (untreated)
1% lethal (treated)
gram negative
coccobacilli
Production Strength
3.5 x 10¹¹ org/ml
2 to 5 day rate-of-action
weeks to months duration-of-action
median infective dose (ID50)
100 org/person

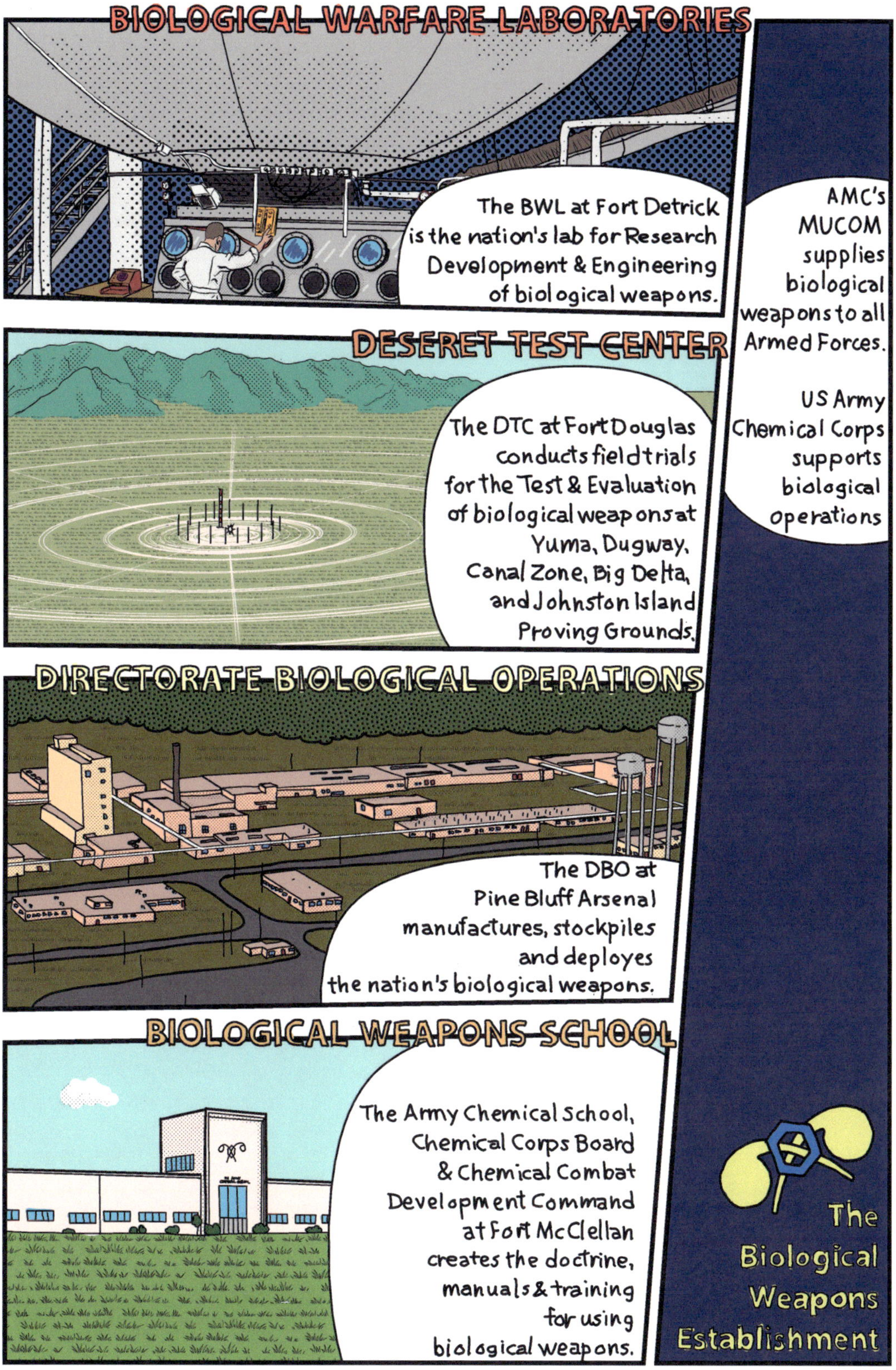
BIOLOGICAL WARFARE LABORATORIES
The BWL at Fort Detrick is the nation's lab for Research Development & Engineering of biological weapons.
DESERET TEST CENTER
The DTC at Fort Douglas conducts field trials for the Test & Evaluation of biological weapons at Yuma, Dugway, Canal Zone, Big Delta, and Johnston Island Proving Grounds.
DIRECTORATE BIOLOGICAL OPERATIONS
The DBO at Pine Bluff Arsenal manufactures, stockpiles and deployes the nation's biological weapons.
BIOLOGICAL WEAPONS SCHOOL
The Army Chemical School, Chemical Corps Board & Chemical Combat Development Command at Fort McClellan creates the doctrine, manuals & training for using biological weapons.
AMC's MUCOM supplies biological weapons to all Armed Forces.
US Army Chemical Corps supports biological operations
The Biological Weapons Establishment

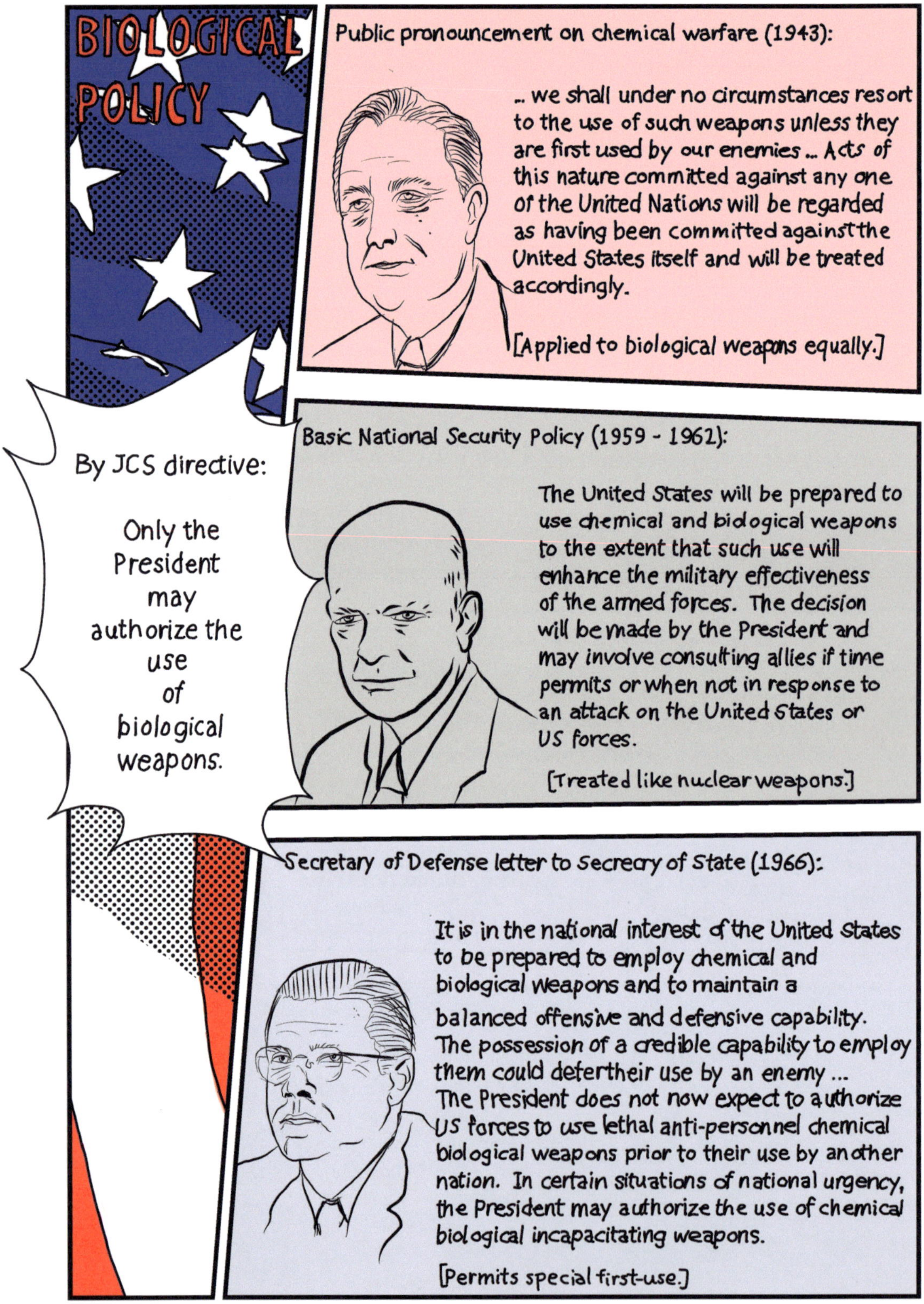
BIOLOGICAL POLICY
Public pronouncement on chemical warfare (1943):
.. we shall under no circumstances resort to the use of such weapons unless they are first used by our enemies ... Acts of this nature committed against any one of the United Nations will be regarded as having been committed against the United States itself and will be treated accordingly.
[Applied to biological weapons equally.]
By JCS directive:
Only the President may authorize the use of biological weapons.
Basic National Security Policy (1959 - 1962):
The United States will be prepared to use chemical and biological weapons to the extent that such use will enhance the military effectiveness of the armed forces. The decision will be made by the President and may involve consulting allies if time permits or when not in response to an attack on the United States or US forces.
[Treated like nuclear weapons.]
Secretary of Defense letter to Secrecry of State (1966):
It is in the national interest of the United States to be prepared to employ chemical and biological weapons and to maintain a balanced offensive and defensive capability. The possession of a credible capability to employ them could defer their use by an enemy ... The President does not now expect to authorize US forces to use lethal anti-personnel chemical biological weapons prior to their use by another nation. In certain situations of national urgency, the President may authorize the use of chemical biological incapacitating weapons.
[Permits special first-use.]

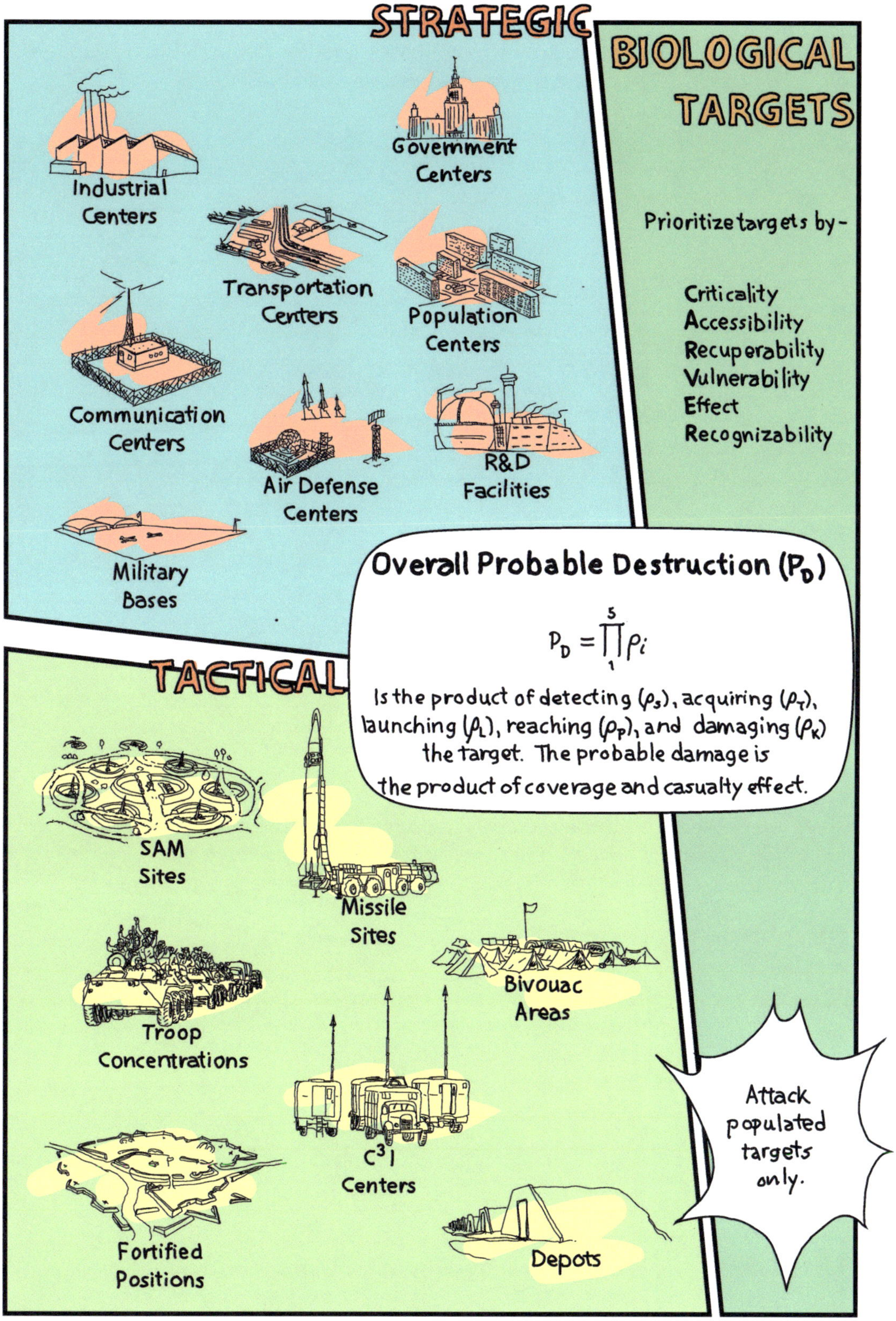
STRATEGIC
BIOLOGICAL TARGETS
Industrial Centers
Government Centers
Transportation Centers
Population Centers
Communication Centers
Air Defense Centers
R&D Facilities
Military Bases
Prioritize targets by -
Criticality
Accessibility
Recuperability
Vulnerability
Effect
Recognizability
Overall Probable Destruction ($P_D$)
$P_D = \prod_{1}^{5} \rho_i$
Is the product of detecting ($\rho_s$), acquiring ($\rho_T$), launching ($\rho_L$), reaching ($\rho_P$), and damaging ($\rho_K$) the target. The probable damage is the product of coverage and casualty effect.
TACTICAL
SAM Sites
Missile Sites
Bivouac Areas
Troop Concentrations
$C^3I$ Centers
Fortified Positions
Depots
Attack populated targets only.

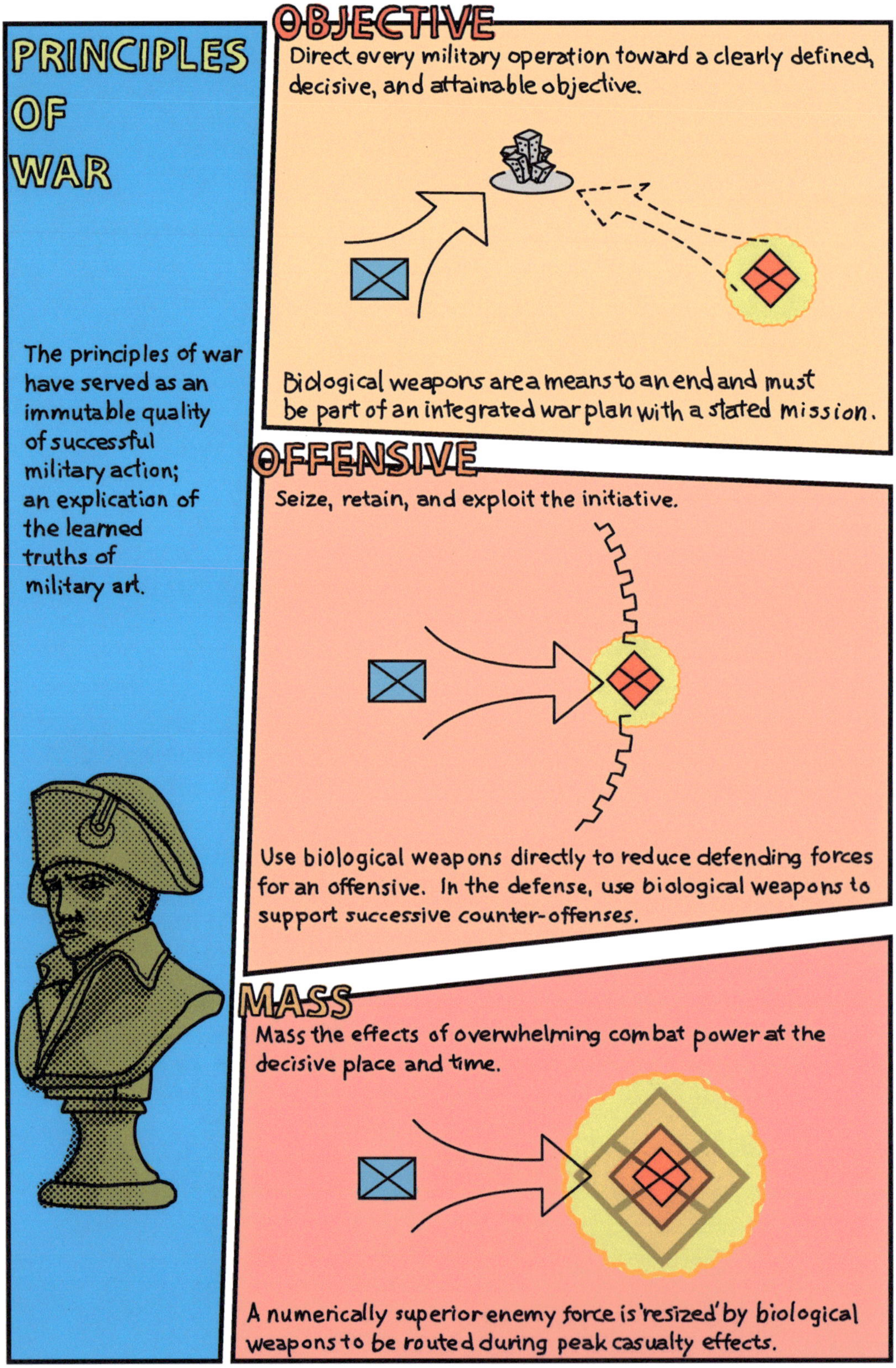
PRINCIPLES OF WAR
The principles of war have served as an immutable quality of successful military action; an explication of the learned truths of military art.
OBJECTIVE
Direct every military operation toward a clearly defined, decisive, and attainable objective.
Biological weapons are a means to an end and must be part of an integrated war plan with a stated mission.
OFFENSIVE
Seize, retain, and exploit the initiative.
Use biological weapons directly to reduce defending forces for an offensive. In the defense, use biological weapons to support successive counter-offenses.
MASS
Mass the effects of overwhelming combat power at the decisive place and time.
A numerically superior enemy force is 'resized' by biological weapons to be routed during peak casualty effects.

## ECONOMY OF FORCE

Employ all combat power available in the most effective way possible; allocate minimum essential combat power to secondary efforts.

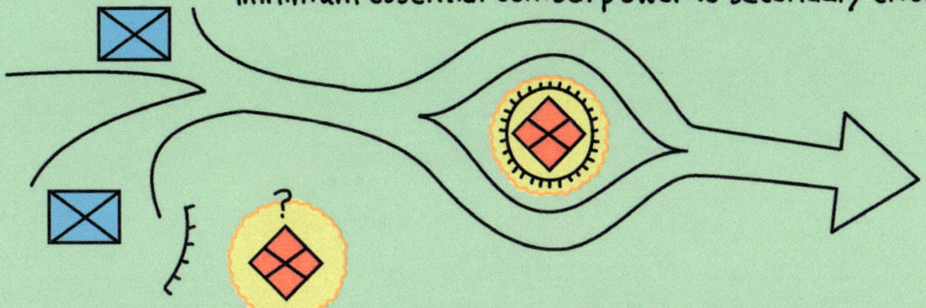

Using biological weapons on bypassed, concealed, or hardened targets will reduce the forces necessary to secure and permit massing forces elsewhere.

## MANEUVER

Place the enemy in a position of disadvantage through the flexible application of combat power.

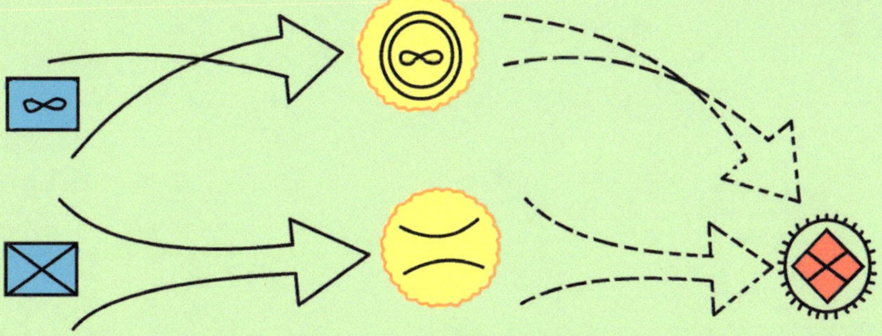

Using biological weapons to destroy enemy forces without physical destruction keeps avenues of advance open and permits the repurposing of captured enemy facilities and material.

## UNITY IN COMMAND

For every objective, seek unity of command and unity of effort.

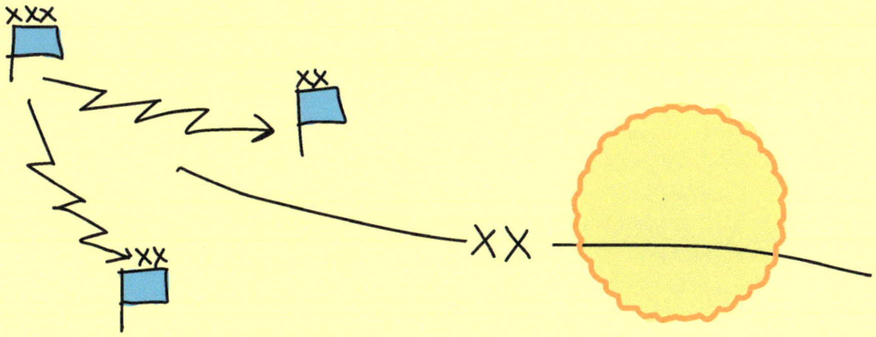


The large area coverage of biological weapons requires coordination by Corps or Army level commands, and should not span across adjacent commands.

## SECURITY

Never permit the enemy to acquire unexpected advantages.

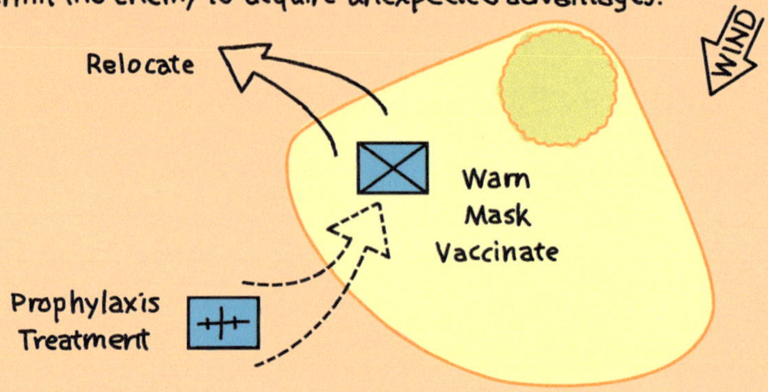


Always consider the off-target dimensions of biological weapons and impacts to troop safety, being prepared to take necessary protective action.

## SURPRISE

Strike the enemy at a time and/or place and in a manner for which he is unprepared.

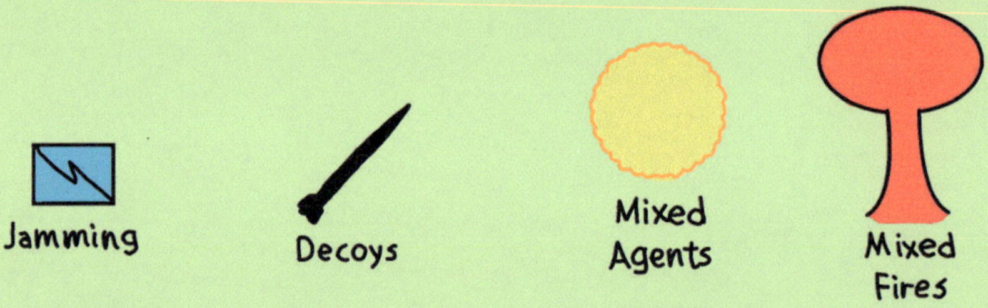


Biological weapons can be defeated by masking, vaccination, and other medical countermeasures. Use biological weapons suddenly and without warning, and in a manner that confounds enemy response to the situation.

## SIMPLICITY

Prepare clear, uncomplicated plans and concise orders to ensure thorough understanding.

Biological weapons have inherent complexities that challenge integration into war plans. Limit the elasticity of weapon effects estimates by using low-end figures in planning, and treating field conditions as a bonus.

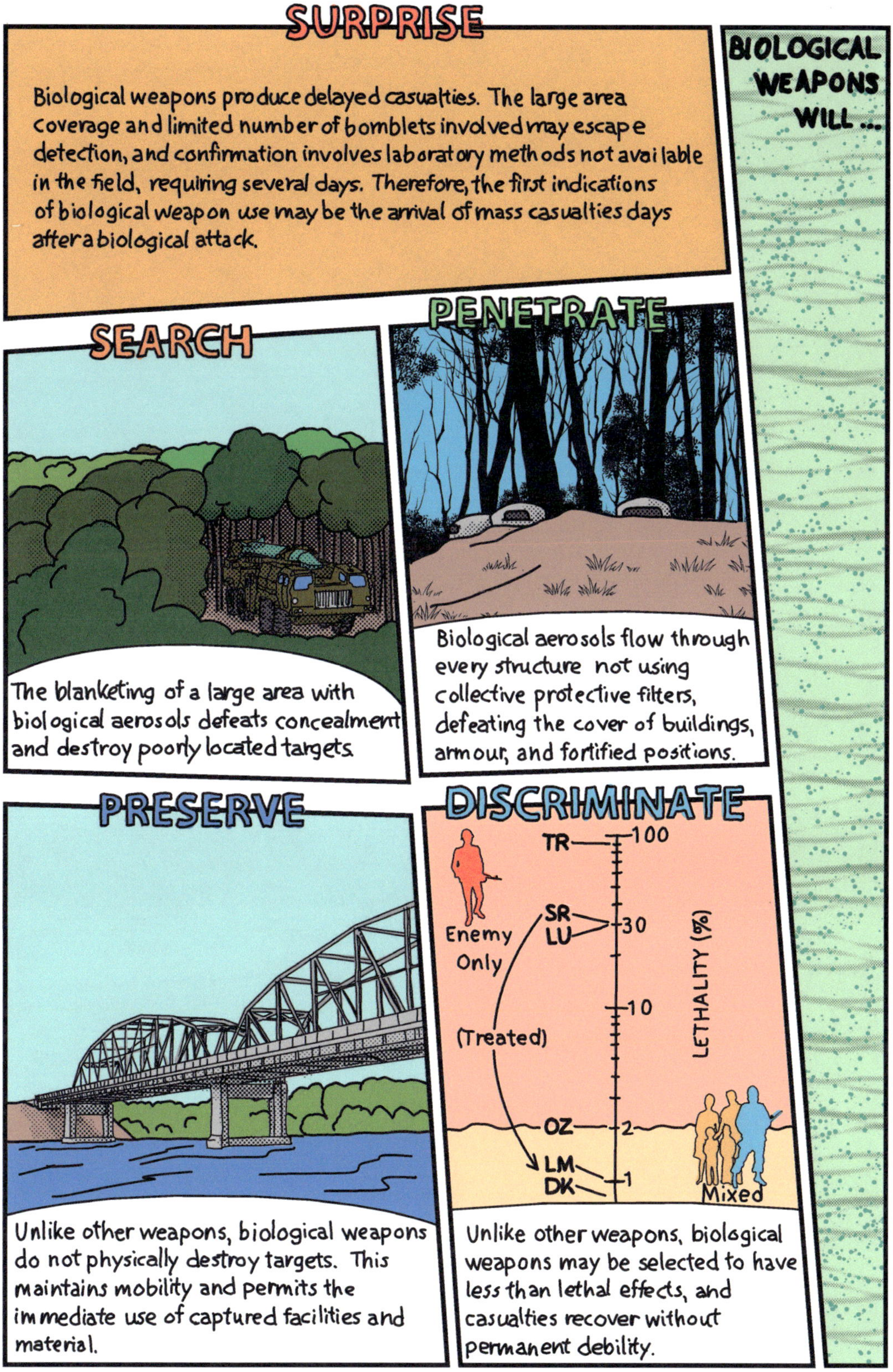
BIOLOGICAL WEAPONS WILL ...
SURPRISE
Biological weapons produce delayed casualties. The large area coverage and limited number of bomblets involved may escape detection, and confirmation involves laboratory methods not available in the field, requiring several days. Therefore, the first indications of biological weapon use may be the arrival of mass casualties days after a biological attack.
SEARCH
The blanketing of a large area with biological aerosols defeats concealment and destroy poorly located targets.
PENETRATE
Biological aerosols flow through every structure not using collective protective filters, defeating the cover of buildings, armour, and fortified positions.
PRESERVE
Unlike other weapons, biological weapons do not physically destroy targets. This maintains mobility and permits the immediate use of captured facilities and material.
DISCRIMINATE
TR
SR
LU
OZ
LM
DK
100
30
10
2
1
LETHALITY (%)
Enemy Only
(Treated)
Mixed
Unlike other weapons, biological weapons may be selected to have less than lethal effects, and casualties recover without permanent debility.

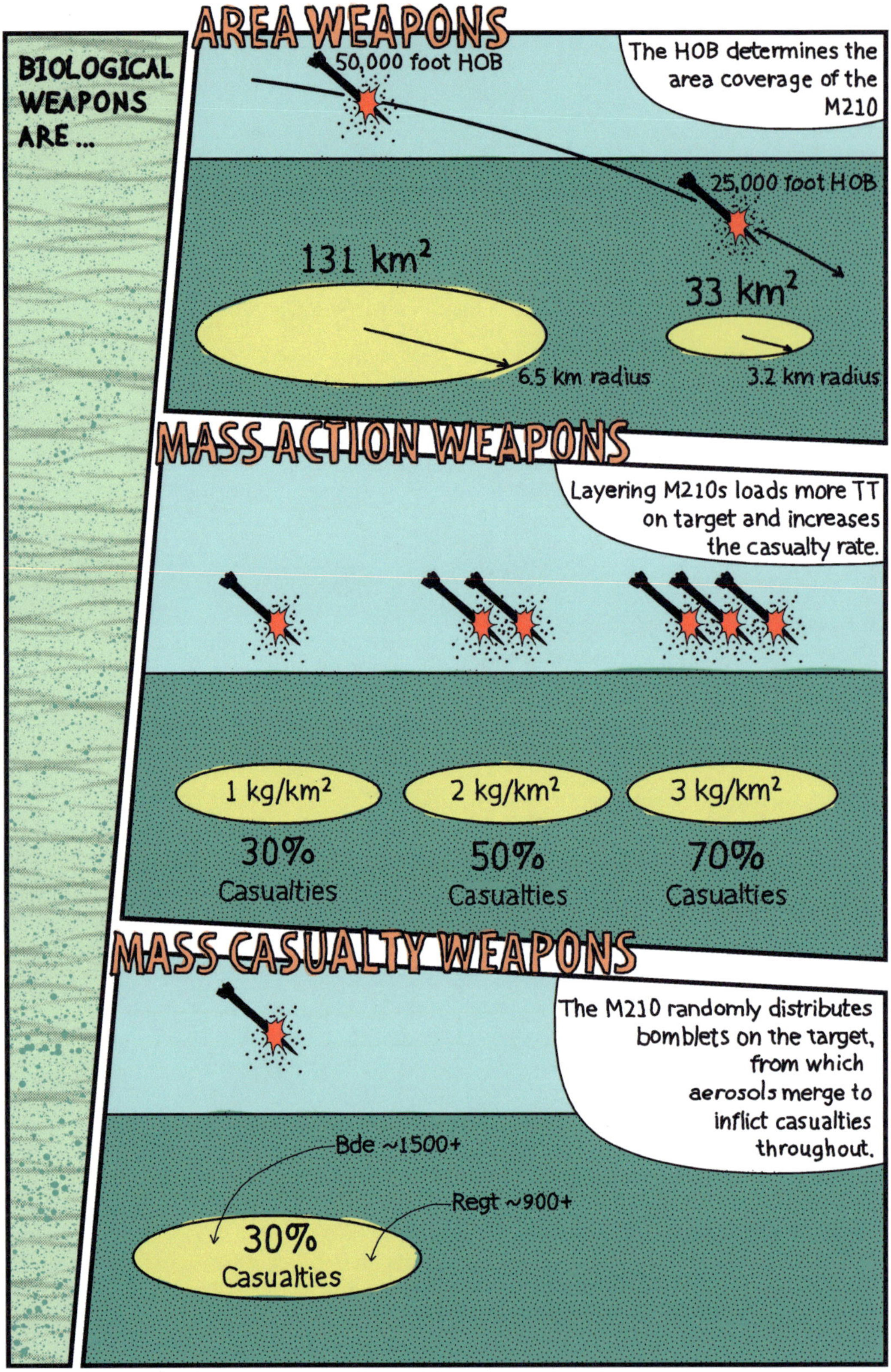
BIOLOGICAL WEAPONS ARE ...
AREA WEAPONS
50,000 foot HOB
The HOB determines the area coverage of the M210
25,000 foot HOB
131 km²
6.5 km radius
33 km²
3.2 km radius
MASS ACTION WEAPONS
Layering M210s loads more TT on target and increases the casualty rate.
1 kg/km²
30% Casualties
2 kg/km²
50% Casualties
3 kg/km²
70% Casualties
MASS CASUALTY WEAPONS
The M210 randomly distributes bomblets on the target, from which aerosols merge to inflict casualties throughout.
Bde ~1500+
Regt ~900+
30% Casualties

## QUANTITY

The amount of biological agent dispersed over an area to inflict casualties —

influenced by number of bomblets, source strength of each bomblet, size of area covered, type of terrain, atmospheric stability, wind speed, and aerobiological decay rate.

(solved graphically)

$Q_0$

$Q_{50}$

($kg/km^2$)

## DOSAGE

The product of concentration and time to biological aerosols to produce casualties -

influenced by rate of breathing and masking.

| Breathing Rate (l/min) | Activity | Adjustment Factor |
|---|---|---|
| 6.5 | Sleeping | 0.6 |
| 10 | Standing | 1.0 (Index) |
| 25 | Walking (3 MPH) | 1.7 |
| 40 | Running (6 MPH) | 3.1 |
| 60 | Fast Run (9 MPH) | 4.8 |

$D_0$

$ICt_{50}$

($HU\text{-}min/m^3$)

## DOSE

The amount of organisms (org) inhaled to produce a casualty —

influenced by aerosol size, susceptibility, and immune system health.

$d_0$

$ID_{50}$

(org/person)

HU (Human Units) is a unit of measure in terms of human median infective doses (i.e., org/$ID_{50}$).

## POSOLOGY

The theory behind chemical biological weapons is the surjective and transitive relation between Quantity, Dosage, and Dose.

$Q_0 \rightarrow D_0 \rightarrow d_0$

$Q_0 \propto D_0 \propto d_0$

The corollary is Weapon Expenditure ($W_0$)

$W_{50}$

($bomblets/km^2$)

## BIOLOGICAL AEROSOLS

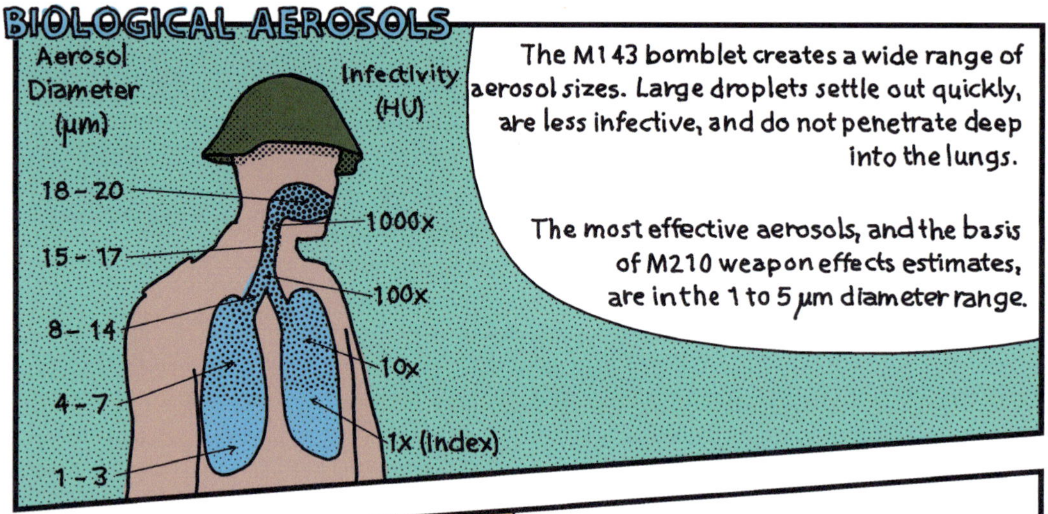


## AGED AEROSOLS

| | Age (hrs) | $ID_{50}$ (org/person) | 5 knot Distance (km) |
|---|---|---|---|
| | 0 | 10 | 0 |
| (Index) | 1 | 100 | 10 |
| | 2 | 250 | 20 |
| | 3 | 3000 | 30 |

## DOSE - RESPONSE

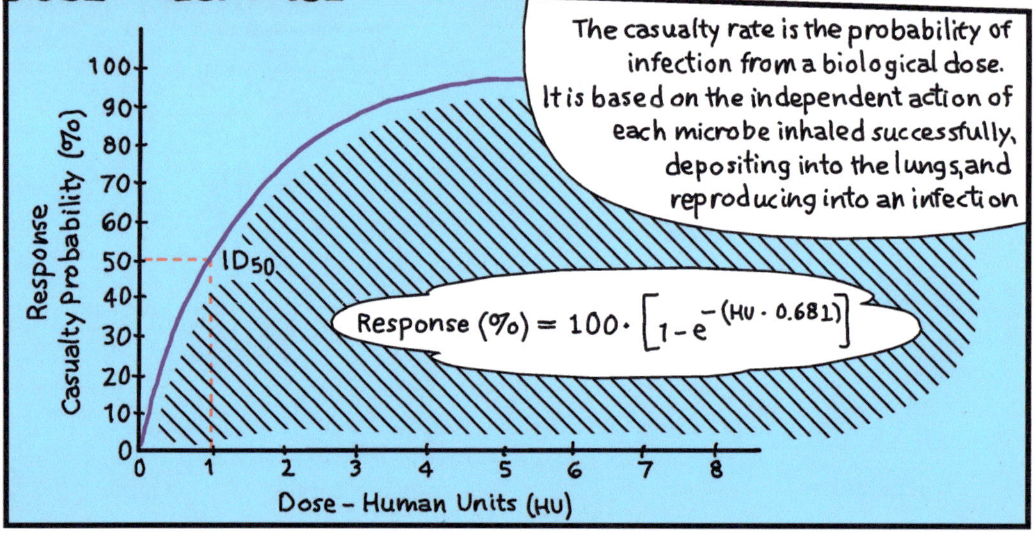

## BIOLOGICAL CASUALTIES

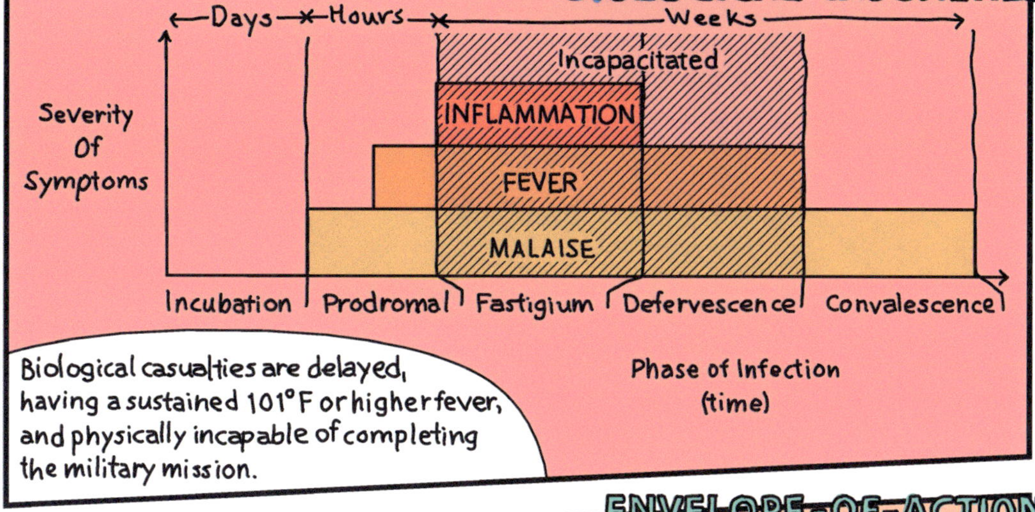


## ENVELOPE-OF-ACTION

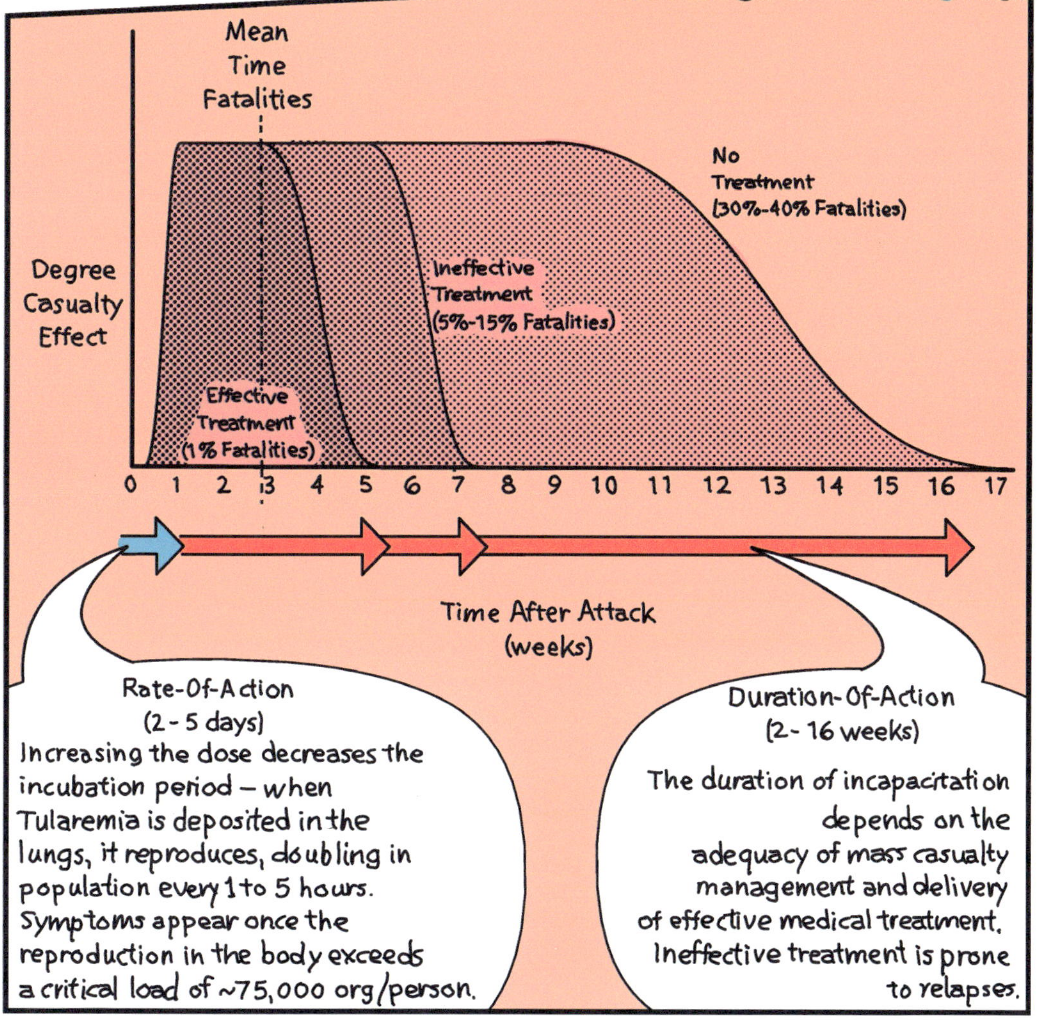

BIOLOGICAL WAR PLANS
FM 3-10
DEPARTMENT OF THE ARMY FIELD MANUAL
CHEMICAL AND BIOLOGICAL WEAPONS EMPLOYMENT
HEADQUARTERS, DEPARTMENT OF THE ARMY
MARCH 1966
Since 1962 Unified Commands are required to review war plans and apply biological weapons in the classified supplement to FM 3-10.
Army or Corps Tactical Operations Center performs biological target analysis.
The CBRE coordinates G2 (weapon effects), G3 (target selection), G4 (SASP placement), and Fire Support Element (Sergeant Bn placement).
TOP SECRET
ANNEX N
BIOLOGICAL WAR PLAN
TOP SECRET
This Page Blank
Unified Commands add Biological Annex to impacted Operational Plans (OPLANs) and Contingency Plans (CONPLANs).
DBO
DBO deploys biological weapons in support of war plans approved by Joint Chiefs of Staff (JCS).

# BIOLOGICAL TARGET ANALYSIS

**1. MISSION**
Command's vision of how M210 will be used.

**4. EVALUATION**
Agent Selection
Type of Attack
Timing
Casualty Rates
Envelope-of-Action
Retroactive Hazards

**2. SITUATION**
Weather
Terrain
Target Selection
Influencing Factors

**3. ANALYSIS**
Capabilities
Counter-capabilities

**5. RECOMMENDATIONS**
Scheduled Fires
On-Call Fires
Troop Safety

# WEAPON EFFECTS ESTIMATES

Biological weapon effects are estimated graphically—

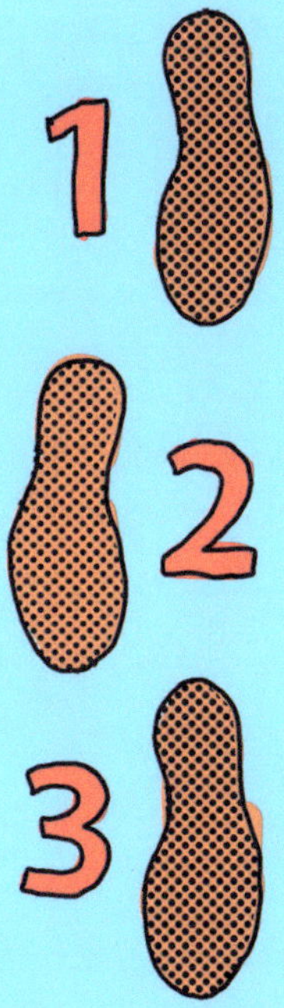


**Amount (Strength)**
Determine the number of median infective doses per a bomblet (HU/bomblet) delivered to target, after estimating number of half-lifes from storage temperature.

**Extent (Area)**
Determine the bomblet density (bomblets/km$^2$) and area of destruction (km$^2$) after finding HOB necessary to cover the target, accounting for delivery and location errors.

**Degree (Casualties)**
Determine the casualty rate (%), after estimating the influence of atmospheric stability, terrain, enemy response, and aerobiological decay rate.

# ATTACK PRODUCTIVITY

There are four elements required for all economic & military productivity. The M210 is for reducing enemy labor. A secondary purpose is to disrupt. When loaded with UT it can deny use of land.

## DISRUPT

Biological strikes, even if of low casualty effect (~10%), produce disproportionate responses when detected, and create uncertainties, increasing the complexity of decision and limiting effective leadership.

## DENY

A corollary to disrupt, the persistent effect of agent-vector combinations renders terrain and facilities uninhabitable for weeks to months, diverting resources to eliminate hazard.

## DESTROY

The M210 is primarily for inflicting mass casualties of 30% or more to reduce labor to a point of halting operations for a set period of time, or until replaced.

# CASUALTY RATE

The required casualty rate depends on degree of labor specialization, overall unit size, and activity.

90% Destroy critical autonomous small units

70% Destroy critical targets

50% Destroy defending units

30% Destroy assualting units or economic centers

**RULE** - DO NOT subject Friendly or Noncombatants to agents with lethalities over 2%.

# TIME-SPACE ASPECTS

The M210 is suited for attacking Regt, Bde, and Div targets with M143 bomblets, extending to Div and Corps targets with future bomblets. Targets of these sizes require duration-of-actions in weeks to defeat enemy operations.

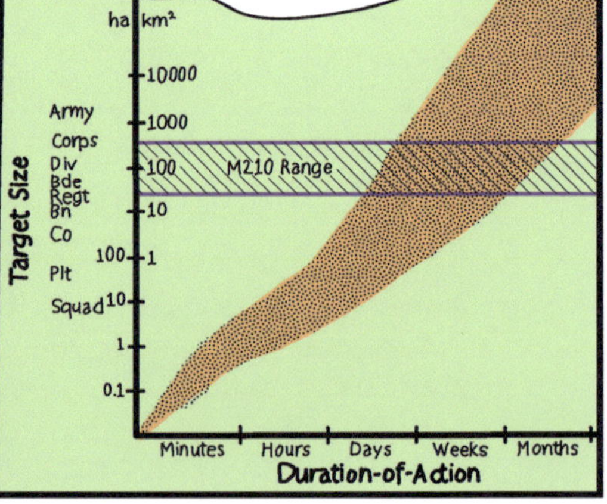

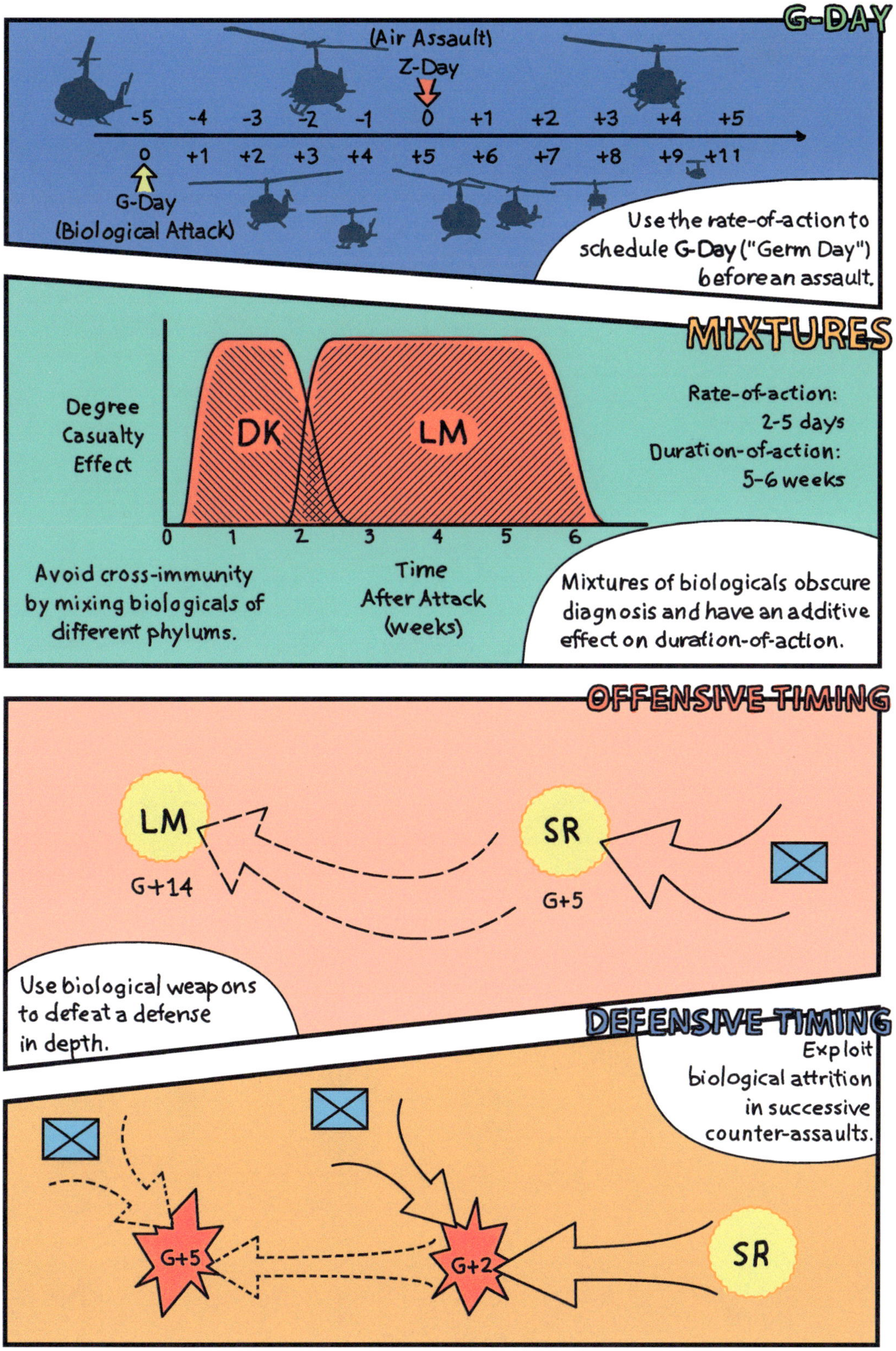
G-DAY
(Air Assault)
Z-Day
-5 -4 -3 -2 -1 0 +1 +2 +3 +4 +5
0 +1 +2 +3 +4 +5 +6 +7 +8 +9 +11
G-Day
(Biological Attack)
Use the rate-of-action to schedule G-Day ("Germ Day") before an assault.
MIXTURES
Degree Casualty Effect
DK
LM
0 1 2 3 4 5 6
Time After Attack (weeks)
Rate-of-action: 2-5 days
Duration-of-action: 5-6 weeks
Avoid cross-immunity by mixing biologicals of different phylums.
Mixtures of biologicals obscure diagnosis and have an additive effect on duration-of-action.
OFFENSIVE TIMING
LM
G+14
SR
G+5
Use biological weapons to defeat a defense in depth.
DEFENSIVE TIMING
Exploit biological attrition in successive counter-assaults.
G+5
G+2
SR

# REVOLVING STOCKPILE

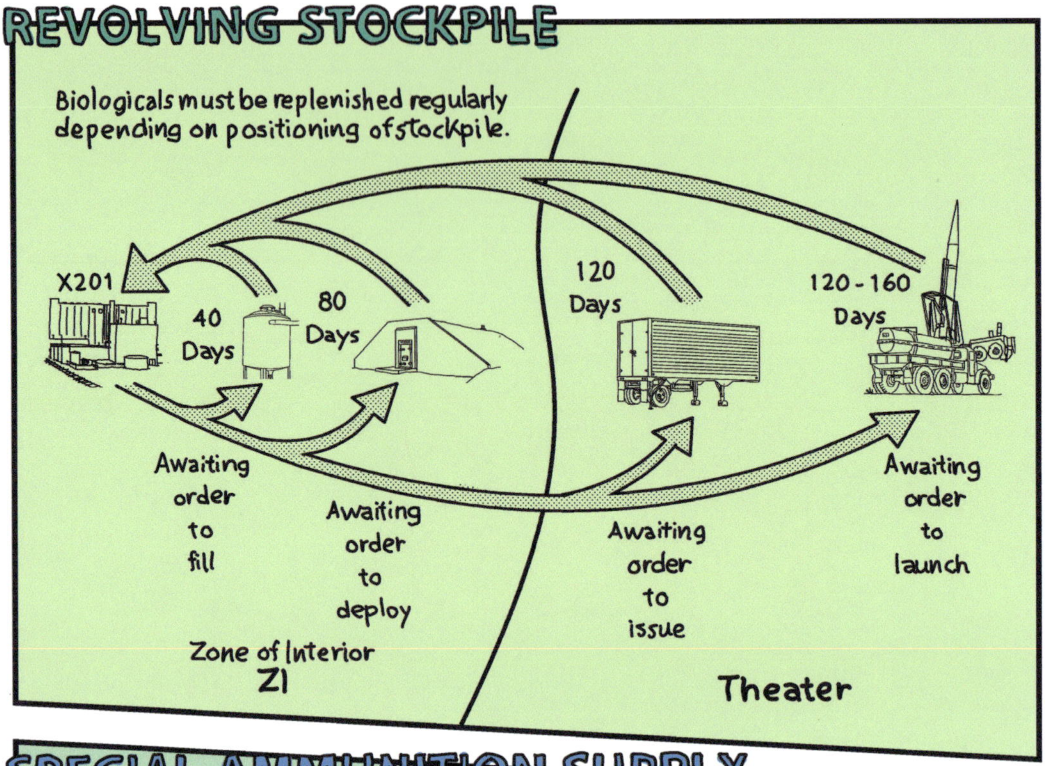


# SPECIAL AMMUNITION SUPPLY

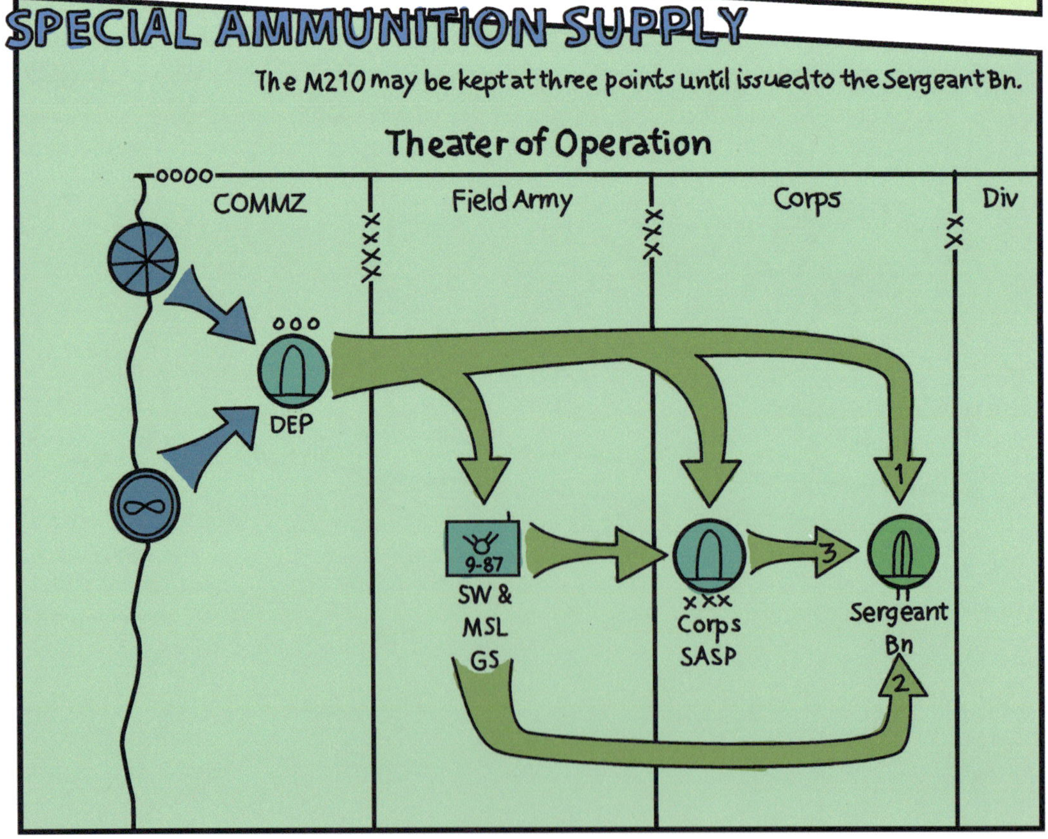

PRODUCTION
The X201 facility uses its 4000 gallon fermenters to produce 400 gallon batches of concentrated and stabilized wet-type Tularemia at a strength of $3.5 \times 10^{11}$ org/ml
(1 batch ≈ 10 M210 warheads)
The DBO runs continuously in stand-by mode at 10% capacity
SHORT-TERM STORAGE
Each batch is kept in holding tanks awaiting orders to fill weapons
The tanks are pasteurized and flushed if not used after one half-life (40 days)
Tanks kept at 41°F (+/-2°F) (densest point of water) to preserve strength
LONG TERM STORAGE
#1005
Bubbling wet-type Tularemia through liquid Freon™ produces frozen pellets that may be stored for 2 to 3 years
These are kept in the DBO Igloo Area at 0°F (freezing point of brine) and thawed to fill weapons
STOCKPILE TO TARGET SEQUENCE
Production to deployment sequence

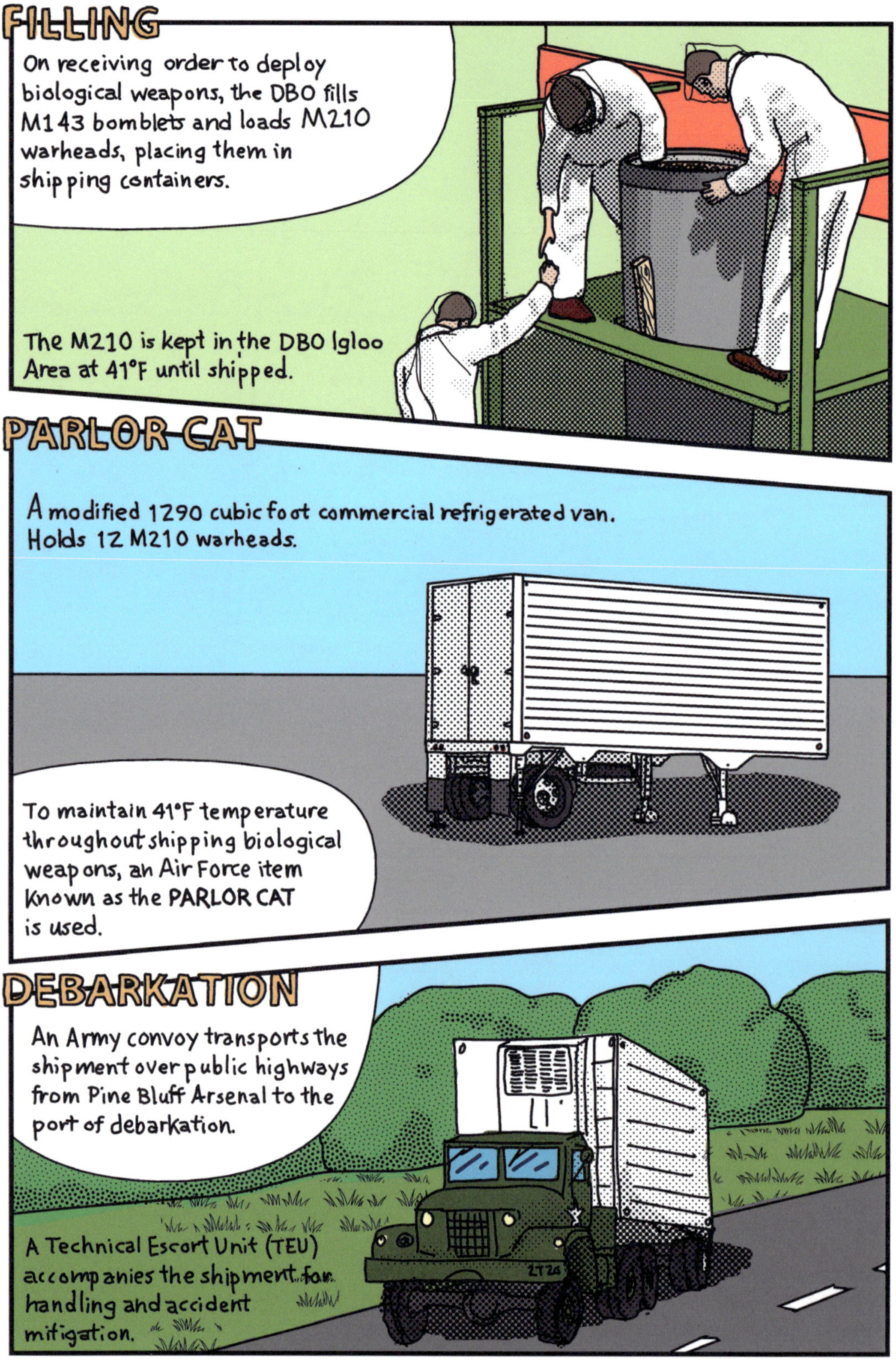
FILLING
On receiving order to deploy biological weapons, the DBO fills M143 bomblets and loads M210 warheads, placing them in shipping containers.
The M210 is kept in the DBO Igloo Area at 41°F until shipped.
PARLOR CAT
A modified 1290 cubic foot commercial refrigerated van. Holds 12 M210 warheads.
To maintain 41°F temperature throughout shipping biological weapons, an Air Force item known as the PARLOR CAT is used.
DEBARKATION
An Army convoy transports the shipment over public highways from Pine Bluff Arsenal to the port of debarkation.
A Technical Escort Unit (TEU) accompanies the shipment for handling and accident mitigation.

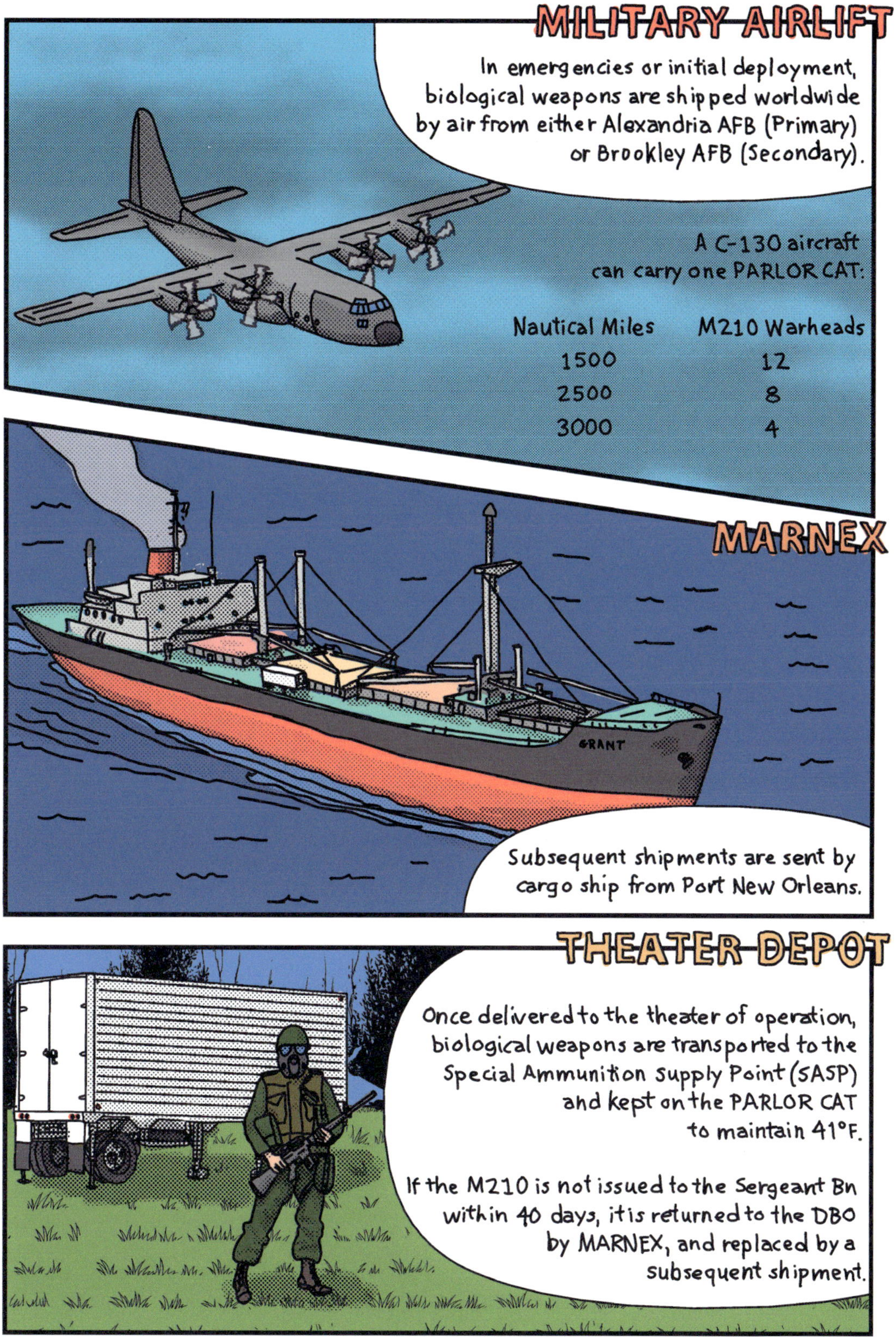
MILITARY AIRLIFT
In emergencies or initial deployment, biological weapons are shipped worldwide by air from either Alexandria AFB (Primary) or Brookley AFB (Secondary).
A C-130 aircraft can carry one PARLOR CAT:
Nautical Miles M210 Warheads
1500 12
2500 8
3000 4
MARNEX
GRANT
Subsequent shipments are sent by cargo ship from Port New Orleans.
THEATER DEPOT
Once delivered to the theater of operation, biological weapons are transported to the Special Ammunition Supply Point (SASP) and kept on the PARLOR CAT to maintain 41°F.
If the M210 is not issued to the Sergeant Bn within 40 days, it is returned to the DBO by MARNEX, and replaced by a subsequent shipment.

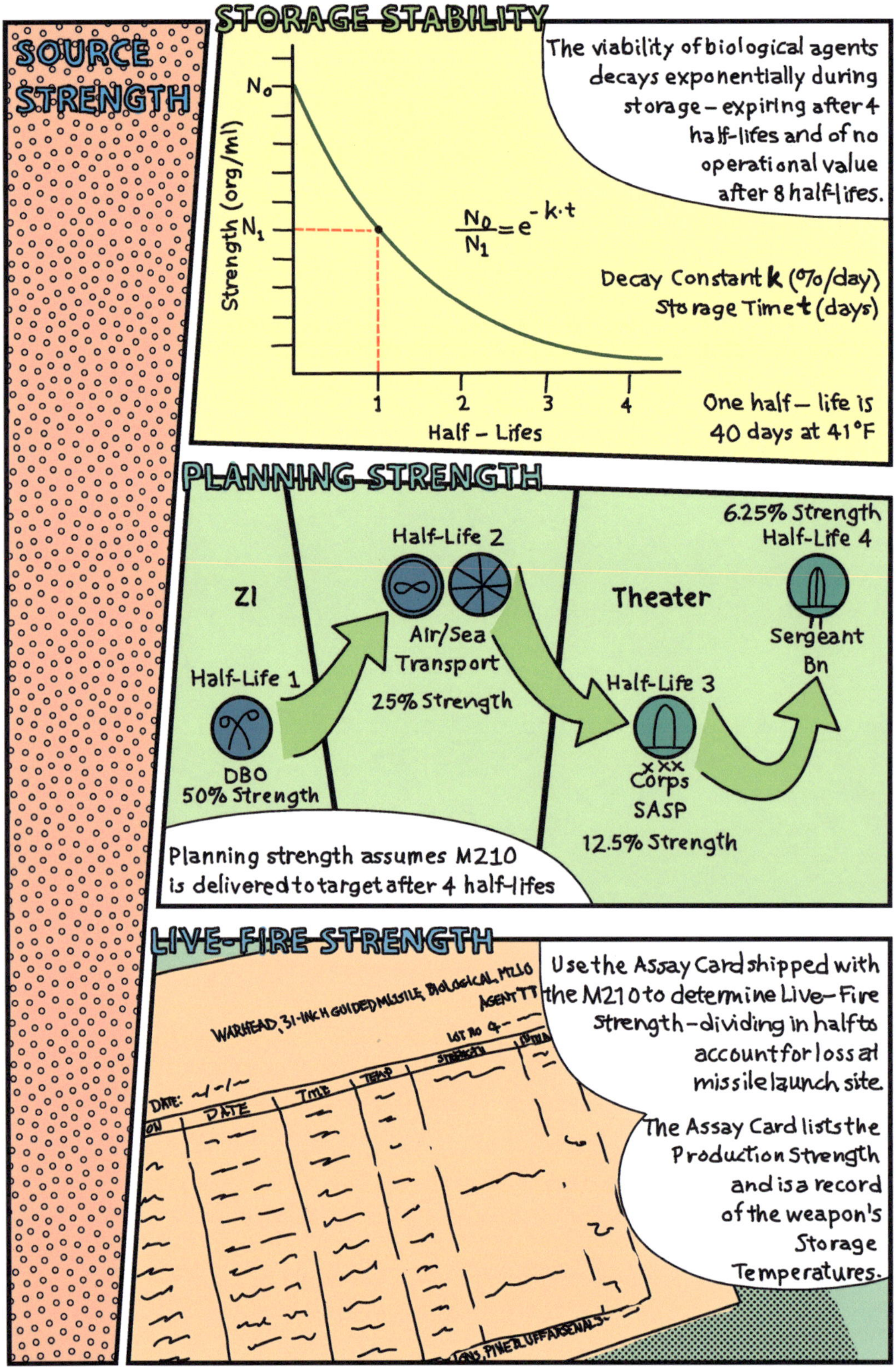
SOURCE STRENGTH
STORAGE STABILITY
The viability of biological agents decays exponentially during storage – expiring after 4 half-lifes and of no operational value after 8 half-lifes.
$N_0$
$N_1$
Strength (org/ml)
$\frac{N_0}{N_1} = e^{-k \cdot t}$
Decay Constant **k** (%/day)
Storage Time **t** (days)
1
2
3
4
Half – Lifes
One half – life is 40 days at 41°F
PLANNING STRENGTH
ZI
Half-Life 1
DBO
50% Strength
Half-Life 2
Air/Sea Transport
25% Strength
Theater
Half-Life 3
Corps
SASP
12.5% Strength
6.25% Strength
Half-Life 4
Sergeant Bn
Planning strength assumes M210 is delivered to target after 4 half-lifes
LIVE-FIRE STRENGTH
Use the Assay Card shipped with the M210 to determine Live-Fire Strength – dividing in half to account for loss at missile launch site.
The Assay Card lists the Production Strength and is a record of the weapon's Storage Temperatures.
WARHEAD, 31-INCH GUIDED MISSILE, BIOLOGICAL, M210
AGENT TT
LOT NO 4-
STRENGTH
DATE
TIME
TEMP
PINE BLUFF ARSENAL

## HALF-LIFE ESTIMATE

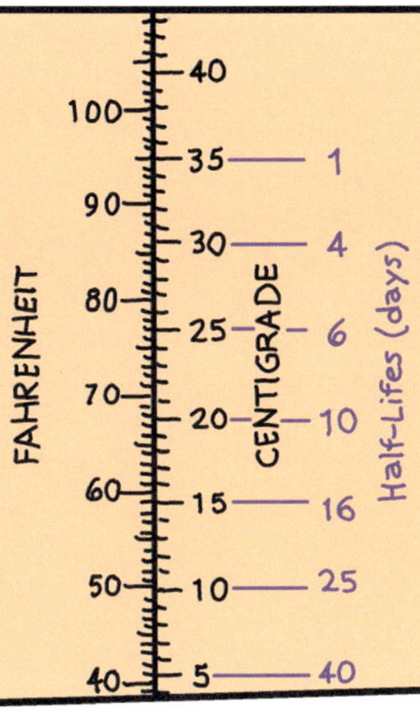


The storage half-life of wet-type Tularemia at its 41° F optimal storage temperature is 40 days. At warmer temperatures the half-life is significantly reduced

| | Temp (°F) | Days | | Half-Life (days) | | |
|---|---|---|---|---|---|---|
| Parlor Cat | 41: | 60 | ÷ | 40 | = | 1.5 |
| M35 Truck | 50: | 5 | ÷ | 25 | = | 0.2 |
| M504 Launcher | 80: | 2 | ÷ | 4 | = | 0.5 |

STTS Total Half-Lifes = 2.2

## FREEZE-THAW LOSS

Wet-type formulations of vegetative microbes like Tularemia are physically destroyed by the shrink-swell of freeze-thaw cycles.

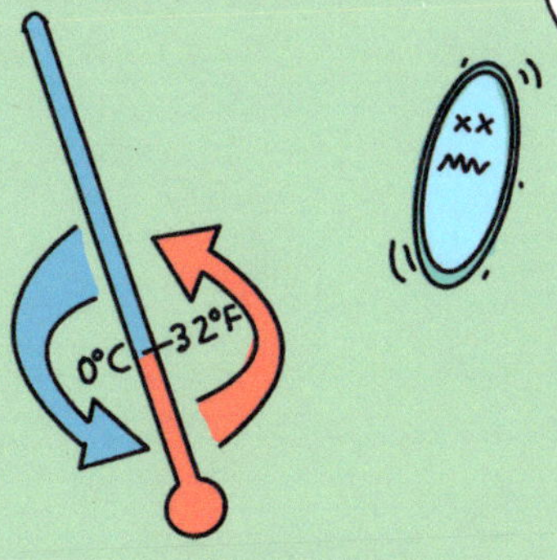


Protect the M210 from freezing temperatures and assume one half-life loss for each freeze-thaw cycle.

## SOURCE STRENGTH

Use the Source Strength Graph to estimate the source strength of each M143 bomblet of the M210, based on the total number of half-lifes from stockpile-to-target.

$$\frac{\text{Agent Strength} \times \text{Bomblet Fill} \times \text{Dissemination Efficiency}}{\text{Median Infective Dose}} = \text{Source Strength}$$

Example:

After 2.2 half-lifes the Source Strength is–

$$\frac{8 \times 10^{10} \text{ org/ml} \times 190 \text{ ml/bomblet} \times 0.2\% \text{ Dissemination}}{100 \text{ org/person}} = 3 \times 10^{8} \text{ HU/bomblet}$$

Or, use Source Strength Graph for 2.2 half-lifes to indicate $3 \times 10^8$ HU/bomblet.
(For other biologicals, adjust Agent Strength, Dissemination Efficiency, and $ID_{50}$.)

# SOURCE STRENGTH GRAPH

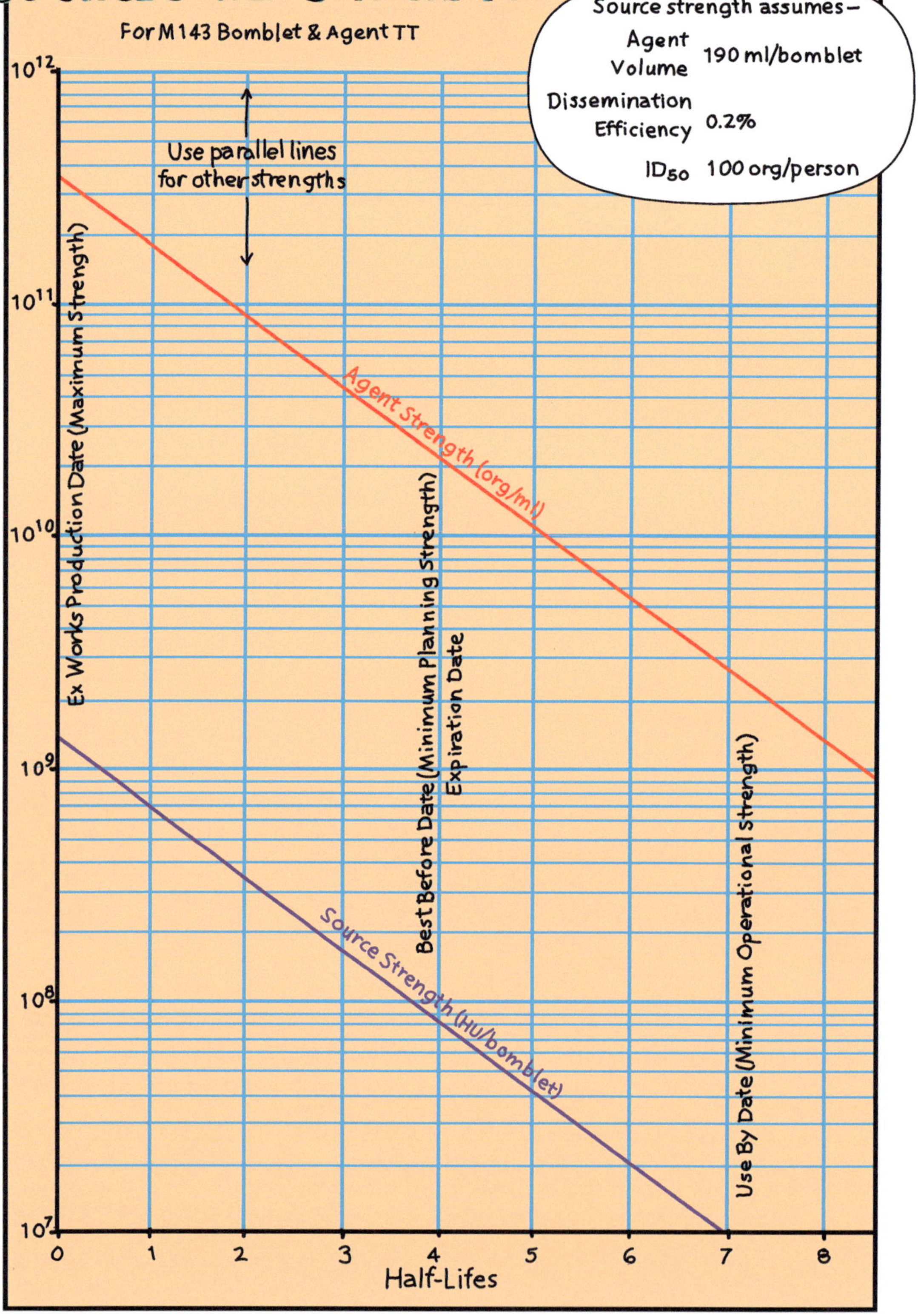

FORWARD SUPPLY
Like nuclear warheads, a helicopter transports the M210 from the SASP to the Sergeant Bn
READINESS STATUS I
The M210 is on missile awaiting launch order
Response Time:
Preplanned fire mission – 15 minutes
Target of Opportunity – 120 minutes
(Warhead must be replaced every few days)
READINESS STATUS II
The M210 is kept at launch site on M35 truck
60 minutes to install warhead
(Add another 25 minutes if replacing a warhead)
(Shipping container sump keeps M210 chilled – must replace every few weeks)
STOCKPILE TO TARGET SEQUENCE
Continued - launch to target sequence.

LAUNCH
At launch the M504 Launcher raises the Sergeant to 75° and rotates to target azimuth.
Launcher either rearms within 83 minutes or relocates to new launch site.
Prelaunch reliability is 78%
PENETRATION
Though the Sergeant is immune to jamming, advanced SAMs are able to intercept, requiring the use of radar jamming and decoys to prevent up to 75% losses.
FLIGHT
The Sergeant is in free-flight after launch. After burnout, drag brakes deploy to control range.
On final descent the guidance system makes two course corrections.
Inflight reliability is 77%

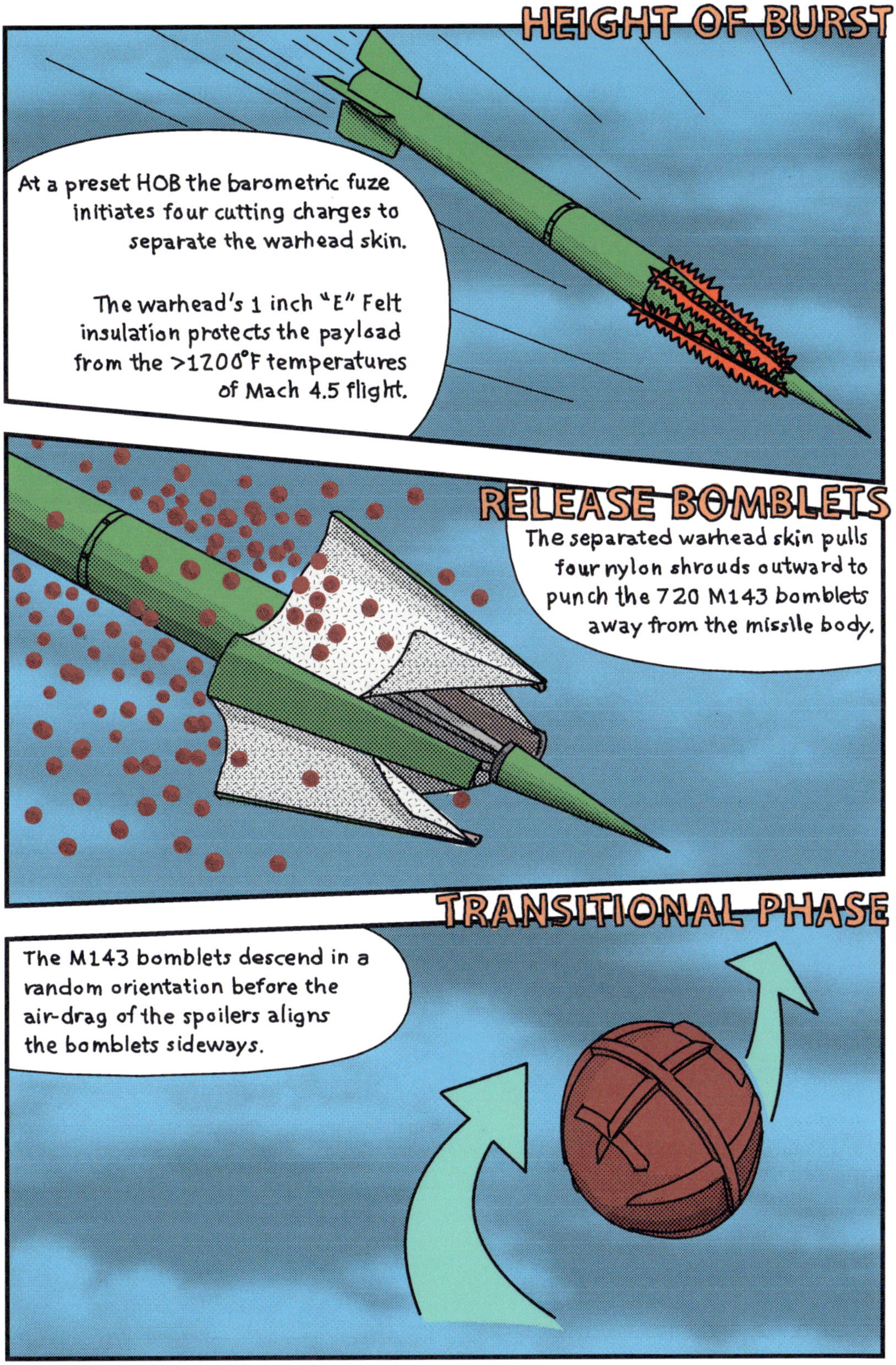
HEIGHT OF BURST
At a preset HOB the barometric fuze initiates four cutting charges to separate the warhead skin.
The warhead's 1 inch "E" Felt insulation protects the payload from the >1200°F temperatures of Mach 4.5 flight.
RELEASE BOMBLETS
The separated warhead skin pulls four nylon shrouds outward to punch the 720 M143 bomblets away from the missile body.
TRANSITIONAL PHASE
The M143 bomblets descend in a random orientation before the air-drag of the spoilers aligns the bomblets sideways.

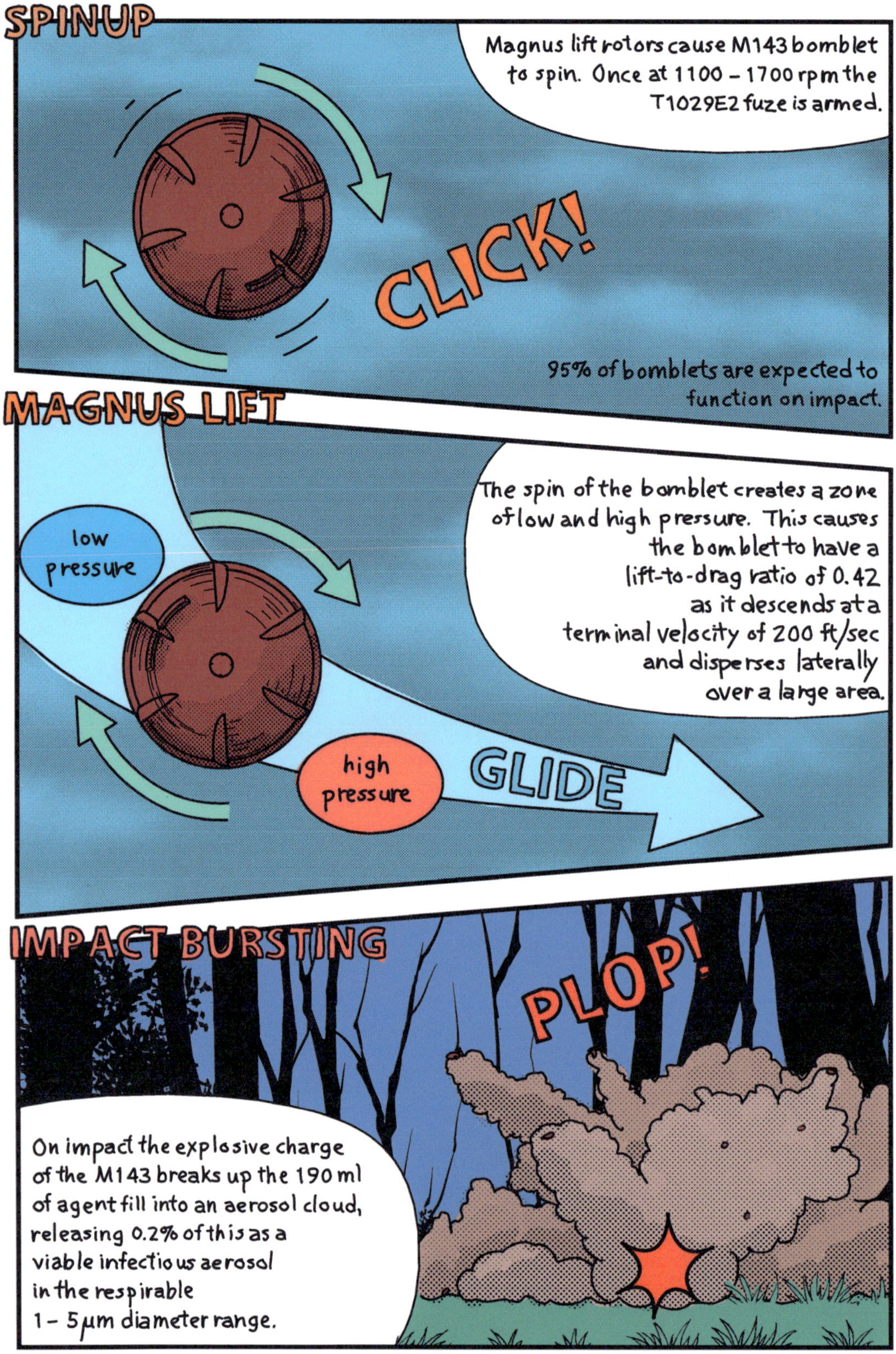
SPINUP
Magnus lift rotors cause M143 bomblet to spin. Once at 1100 – 1700 rpm the T1029E2 fuze is armed.
CLICK!
95% of bomblets are expected to function on impact.
MAGNUS LIFT
The spin of the bomblet creates a zone of low and high pressure. This causes the bomblet to have a lift-to-drag ratio of 0.42 as it descends at a terminal velocity of 200 ft/sec and disperses laterally over a large area.
low pressure
high pressure
GLIDE
IMPACT BURSTING
PLOP!
On impact the explosive charge of the M143 breaks up the 190 ml of agent fill into an aerosol cloud, releasing 0.2% of this as a viable infectious aerosol in the respirable 1 – 5μm diameter range.

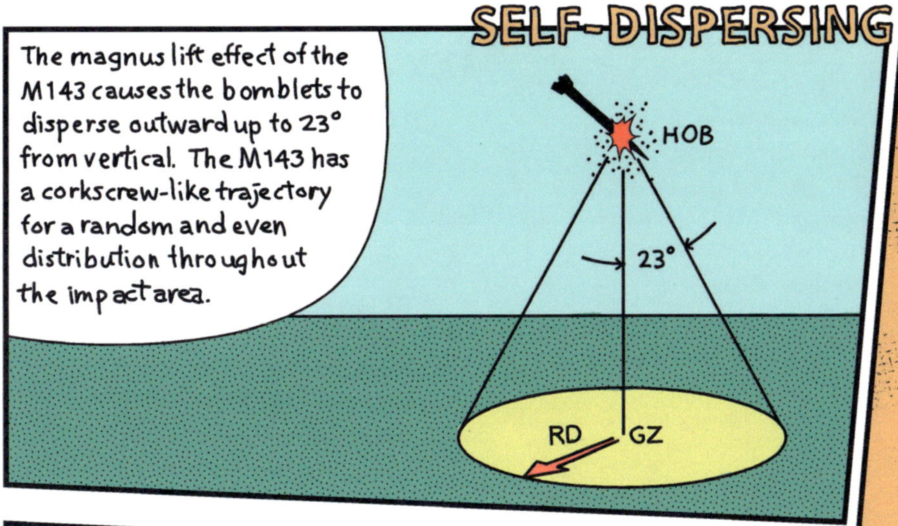


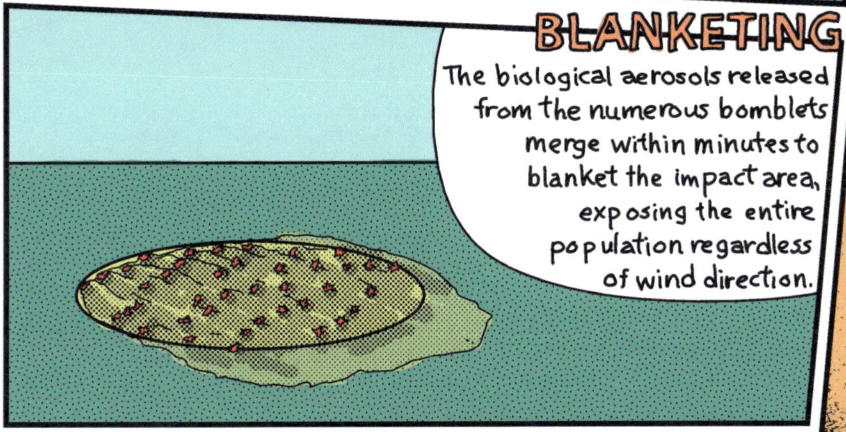


## BOMBLET DENSITY

| HOB (ft) | RD Radius of Destruction (km) | Area of Destruction ($km^2$) | Functioning* Bomblet Density (bomblets/$km^2$) |
|---|---|---|---|
| 25,000 | 3.2 | 33 | 21 |
| 30,000 | 3.9 | 47 | 14 |
| 35,000 | 4.5 | 64 | 11 |
| 40,000 | 5.2 | 84 | 8 |
| 45,000 | 5.8 | 106 | 6 |
| 50,000 | 6.5 | 131 | 5 |

*Assumes each M210 delivers 684 bomblets (95% functioning)

# TOTAL COVERAGE

The Circular Error Probability (CEP) of the Sergeant is 304m, meaning 50% of the missiles will release bomblets within 304m of the Designated Ground Zero (DGZ). The Circular Distribution for 90% of the missiles (or CD90) is 550m. Reducing the Radius of Destruction (RD) by the CD90 indicates the maximum target size that the M210 will inflict casualties on with 90% assurance, accounting for delivery errors.

## MAXIMUM TARGET SIZE

(for 90% assured coverage – no TLE)

| HOB (ft) | Radius (km) | Area ($km^2$) |
|---|---|---|
| 25,000 | 2.7 | 23 |
| 30,000 | 3.3 | 35 |
| 35,000 | 4.0 | 50 |
| 40,000 | 4.6 | 67 |
| 45,000 | 5.3 | 87 |
| 50,000 | 5.9 | 110 |

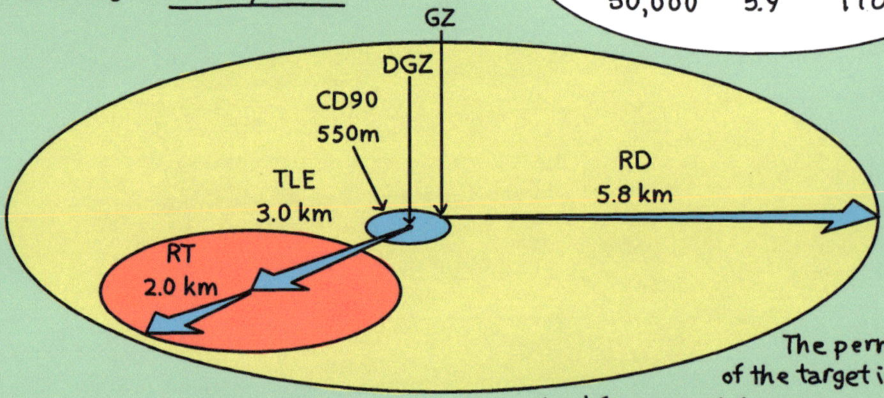


The permanence of the target is also an important factor, and the area covered must account for the Target Location Error (TLE). If a target has a 2 km Radius of Target (RT), and also a 3 km TLE, then the total Radius of Destruction (RD) for assured coverage must be over 5 km to account for location errors.

$$RD \geq RT + TLE + CD90$$

# FRACTIONAL COVERAGE

When the area covered by the M210 is less than the size of the target, use the Area Target Graph to estimate the fraction of target covered. Multiplying the casualty rate of the attack by this fraction of target covered indicates the expected casualty rate to the target.

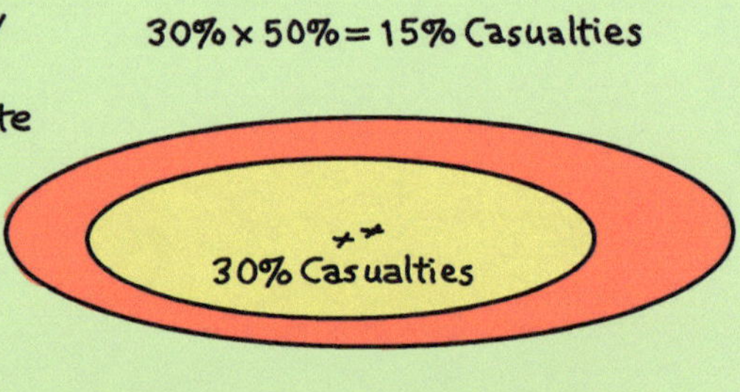


Example: If the Radius of Target (RT) is 6500m, the Radius of Destruction (RD) is 4500m, and the CD90 is 550m, then the fraction of target covered is ~50%.

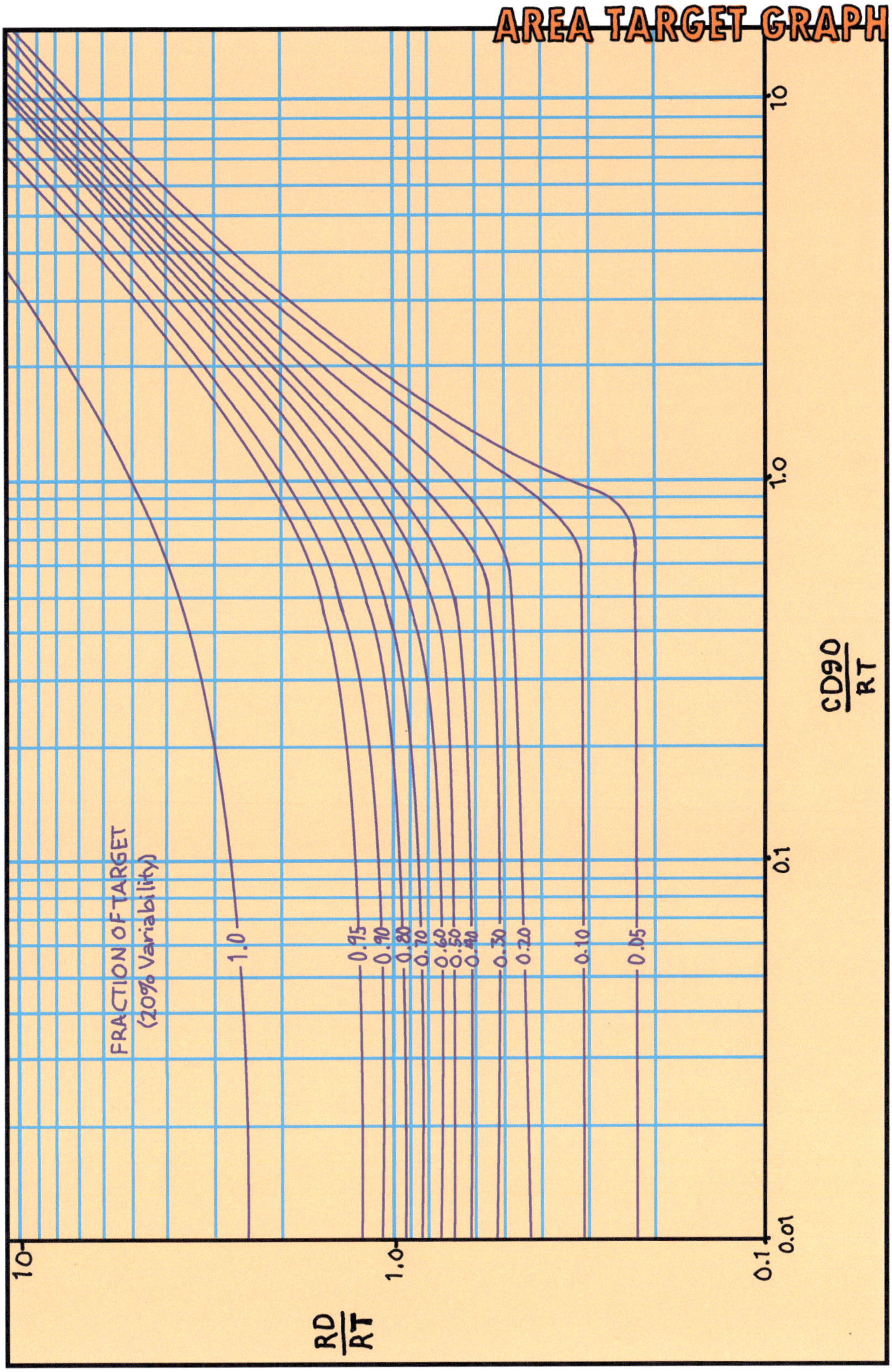
AREA TARGET GRAPH
FRACTION OF TARGET
(20% Variability)
1.0
0.95
0.90
0.80
0.70
0.60
0.50
0.40
0.30
0.20
0.10
0.05
CD90 / RT
0.01
0.1
1.0
10
RD / RT
0.1
1.0
10

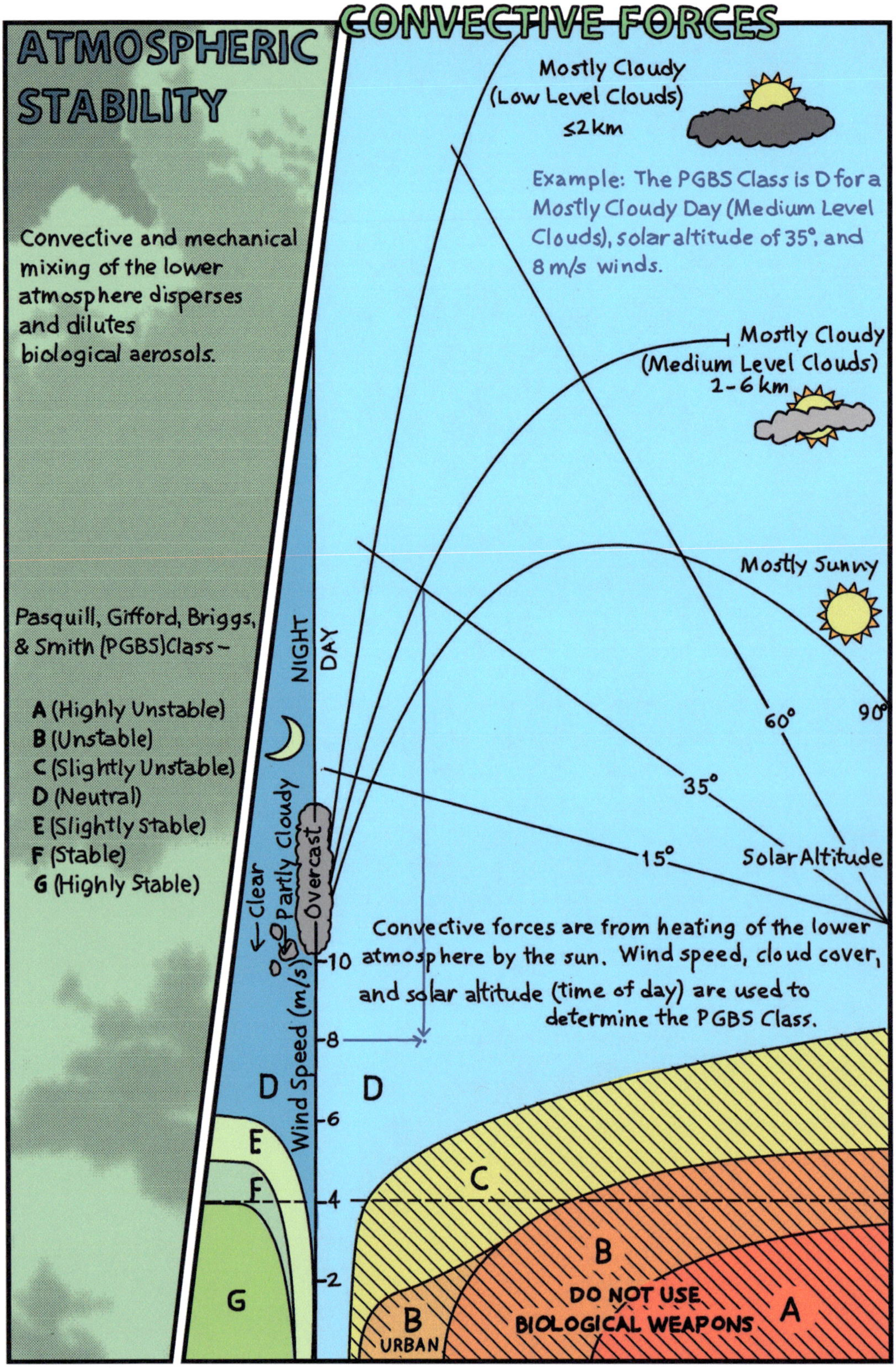
ATMOSPHERIC STABILITY
CONVECTIVE FORCES
Convective and mechanical mixing of the lower atmosphere disperses and dilutes biological aerosols.
Pasquill, Gifford, Briggs, & Smith (PGBS) Class –
A (Highly Unstable)
B (Unstable)
C (Slightly Unstable)
D (Neutral)
E (Slightly Stable)
F (Stable)
G (Highly Stable)
Mostly Cloudy
(Low Level Clouds)
≤2km
Example: The PGBS Class is D for a Mostly Cloudy Day (Medium Level Clouds), solar altitude of 35°, and 8 m/s winds.
Mostly Cloudy
(Medium Level Clouds)
2-6km
Mostly Sunny
NIGHT
DAY
60°
90°
35°
15°
Solar Altitude
Clear
Partly Cloudy
Overcast
Wind Speed (m/s)
10
8
6
4
2
Convective forces are from heating of the lower atmosphere by the sun. Wind speed, cloud cover, and solar altitude (time of day) are used to determine the PGBS Class.
D
D
E
F
G
C
B
B
URBAN
A
DO NOT USE
BIOLOGICAL WEAPONS

## GROUND COVER

Mist, haze, frost, and snow reduce convective forces and increase atmospheric stability.

| | PGBS Class | | |
|---|---|---|---|
| Dry surfaces (visibility >4 km) | B | D | F |
| Wet surfaces (visibility 1 – 4 km) | D | D | F |
| Frost covered (visibility <1 km) | D | F | F |
| Snow covered | F | F | F |

## MECHANICAL FORCES

Terrain roughness breaks up biological aerosols—Adjust source strength on OPEN Weapon Effects Graphs for terrain conditions.

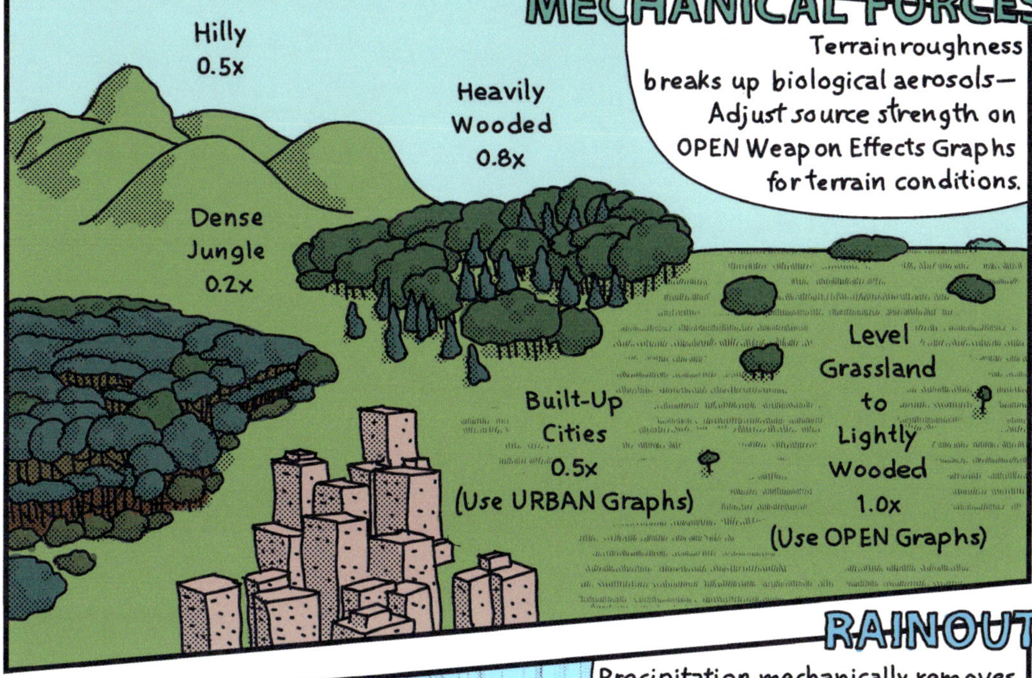


## RAINOUT

Precipitation mechanically removes biological aerosols from the air—Adjust source strength of Neutral (D) stability weapon effects graphs to account for losses from rainout.

| Precipitation | Rate (inches/hour) | Adjustment Factor |
|---|---|---|
| Light | 0.01 – 0.10 | 0.8x |
| Moderate | 0.10 – 0.50 | 0.4x |
| Heavy | 0.50 – 2.00 | 0.2x |

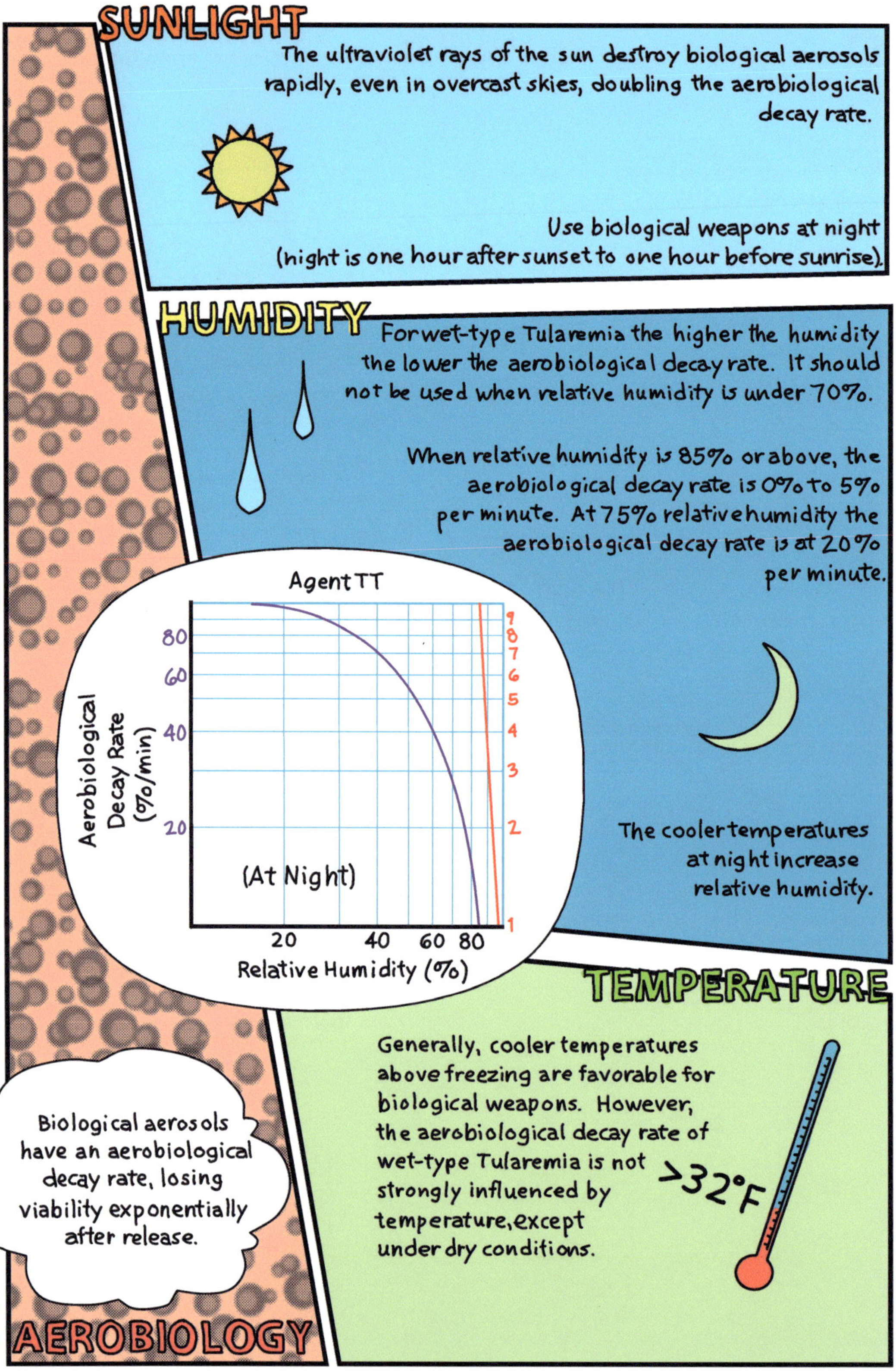
SUNLIGHT
The ultraviolet rays of the sun destroy biological aerosols rapidly, even in overcast skies, doubling the aerobiological decay rate.
Use biological weapons at night (night is one hour after sunset to one hour before sunrise).
HUMIDITY
For wet-type Tularemia the higher the humidity the lower the aerobiological decay rate. It should not be used when relative humidity is under 70%.
When relative humidity is 85% or above, the aerobiological decay rate is 0% to 5% per minute. At 75% relative humidity the aerobiological decay rate is at 20% per minute.
Agent TT
Aerobiological Decay Rate (%/min)
80
60
40
20
9
8
7
6
5
4
3
2
1
(At Night)
20
40
60
80
Relative Humidity (%)
The cooler temperatures at night increase relative humidity.
TEMPERATURE
Generally, cooler temperatures above freezing are favorable for biological weapons. However, the aerobiological decay rate of wet-type Tularemia is not strongly influenced by temperature, except under dry conditions.
>32°F
Biological aerosols have an aerobiological decay rate, losing viability exponentially after release.
AEROBIOLOGY

## MASKING

Masking is not normally a problem with biological attacks as it takes 2 to 3 days to determine from sampling aerosols. Nonetheless, a prepared enemy may take to masking on the first alert of a missile launch. In these cases adjust casualty estimates by the following:

Good Mask Discipline - 0.10x
Poor Mask Discipline - 0.25x

EXAMPLE: A 30% casualty effect on a target with poor mask discipline results in 7.5% casualties.

## VACCINATION

The live vaccine strain of Tularemia used by US forces is ~80% effective. Sino-Soviet forces use a killed Tularemia vaccine which is only ~40% effective.

In the case where enemy populations have been vaccinated, adjust the casualty rate by 0.5x.

EXAMPLE: A 50% casualty effect on a vaccinated target results in 25% casualties.

## TREATMENT

Without medical treatment the duration-of-action of Tularemia is 2 to 4 months with a lethality of 30%.

Sino-Soviet forces use streptomycin as a standard therapy. The Schu S5 strain of Tularemia (SR) is streptomycin resistant. This ineffective therapy will at best reduce lethality to 5% - 15% and is unlikely to reduce the duration-of-action below 4 weeks.

Modern therapies available to US forces can reduce the duration-of-action to 2-4 weeks and lethality to <1%.

The casualty rate remains unchanged, only adjusting the estimate of lethality and expected duration-of-action. If streptomycin therapy is started before onset, adjust casualty rate by 0.5x.

# BIOLOGICAL WEAPON EFFECTS

## ESTIMATES

To achieve the mission, area source graphs are used to estimate—

- WARHEADS – How many weapons are needed?
- CASUALTIES – How many casualties are produced?
- HALF-LIFES – What is the usable age of weapons?

Weapon systems like the M210 that use self-dispersing biological bomblets are area source weapons.

## WEAPON EFFECTS GRAPHS

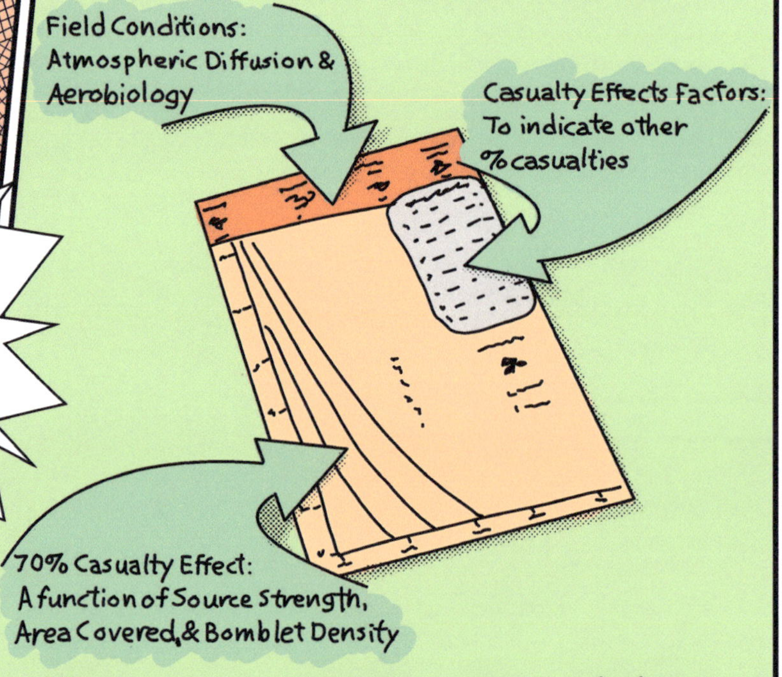


| Weapon Effects Graph | Terrain Complex | Atmospheric Stability (PGBS Class) | Wind Speed (m/s) | Aerobiological Decay Rate (%/min) |
|---|---|---|---|---|
| A | Open | F | <6 | 0-5 |
| B | Open | D | >0 | 0-5 |
| C | Urban | F | <6 | 0-5 |
| D | Urban | D | >0 | 0-5 |
| E | Open/Urban | G-F | <4 | 20 |
| F | Open/Urban | F-D | >4 | 20 |

## SELECTING GRAPH

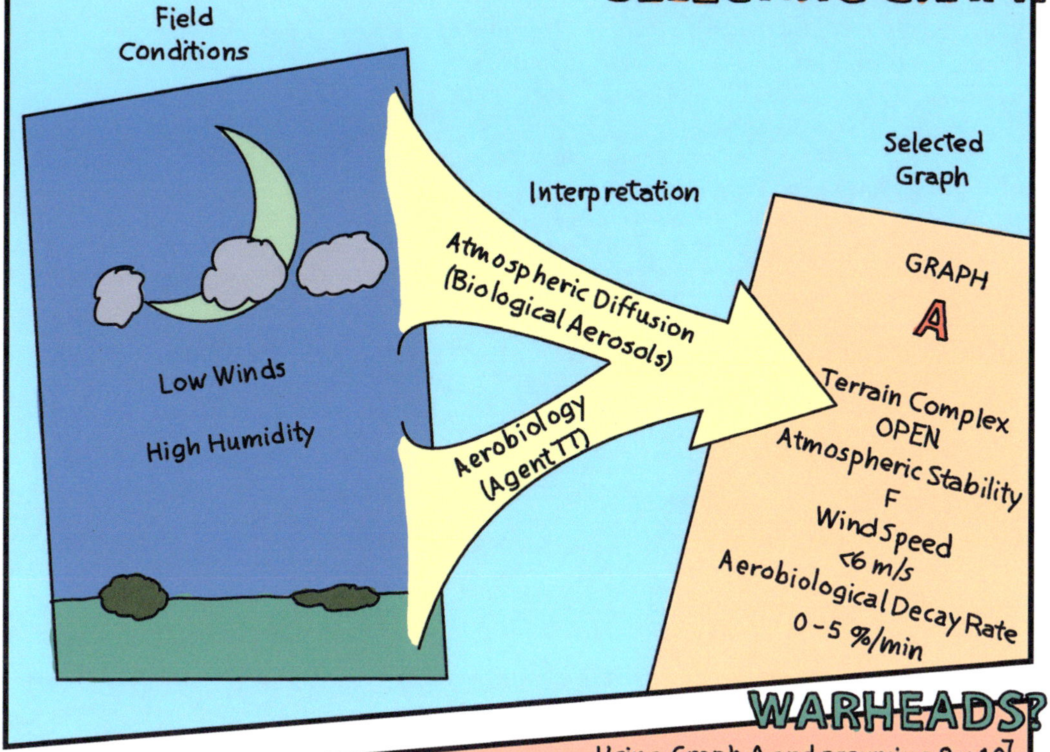


## WARHEADS?

Using Graph A and assuming $8 \times 10^7$ HU/bomblet Source Strength (4 half-lifes), find the number of M210s necessary to inflict 90% casualties to a 131 $km^2$ area.

**STEP 1 Bomblet Density (70% Casualties)**
Using Graph A, the 131 $km^2$ curve intercepts the Source Strength of $8 \times 10^7$ HU/bomblet to indicate 6 bomblets/$km^2$.

**STEP 2 Casualty Effect Factor (90% Casualty Effect)**
Using the Casualty Effect Factor table on Graph A, run down the $10^8$ HU/bomblet (nearest) Source Strength column to find the factor of 4.00.

**STEP 3 Bomblet Density (90% Casualties)**
Multiplying 6 bomblets/$km^2$ by 4.00 indicates 24 bomblets/$km^2$.

**STEP 4 M210s (131 $km^2$ Area of Destruction)**
Dividing 24 bomblets/$km^2$ by the Functioning Bomblet Density of 5 bomblets/$km^2$ for the M210 indicates 4.8 M210s.

**SOLUTION**
Layering 5 M210 warheads on a 131 $km^2$ area will have a 90% Casualty Effect (round to next whole integer).

## CASUALTIES?

Using Graph A and assuming a Source Strength of $8 \times 10^7$ HU/bomblet, find the percent casualties from one M210 warhead on a 131 $km^2$ area.

**STEP 1** Bomblet Density (70% Casualties)
Using Graph A, the 131 $km^2$ curve intercepts the Source Strength of $8 \times 10^7$ HU/bomblet to indicate 6 bomblets/$km^2$.

**STEP 2** Casualty Effect Factor (131 $km^2$ Area of Destruction)
Dividing the Functioning Bomblet Density of 5 bomblets/$km^2$ by 6 bomblets/$km^2$ indicates 0.83.

**STEP 3** Casualty Effect (%)
Using the Casualty Effect Factor table on Graph A, run down the $10^8$ HU/bomblet (nearest) Source Strength column to find the factor of 0.83 to indicate 60%.

**SOLUTION**
One M210 warhead will inflict 60% casualties on a 131 $km^2$ area after four half-lifes.

## HALF-LIFES?

Using Graph A, find the number of maximum half-lifes the M210 can have and still be able to inflict 30% casualties on a 131 $km^2$ area.

**STEP 1** Casualty Effect Factor (30% Casualty Effect)
Using the Casualty Effect Factor table on Graph A, run down the $10^7$ HU/bomblet (lowest) Source Strength column to find the factor of 0.28.

**STEP 2** Bomblet Density (70% Casualties)
Dividing the Functioning Bomblet Density of 5 bomblets/$km^2$ by 0.28 indicates 18 bomblets/$km^2$.

**STEP 3** Source Strength (131 $km^2$ Area of Destruction)
Using Graph A, the 131 $km^2$ curve intercepts the Bomblet Density of 18 bomblets/$km^2$ to indicate $1.2 \times 10^7$ HU/bomblet.

**STEP 4** Half-Lifes (30% Casualties)
Using the Source Strength Graph, $1.2 \times 10^7$ HU/bomblet indicates 6.75.

**SOLUTION**
The M210 must not be older than 6.75 half-lifes if it is to inflict ≥30% casualties on 131 $km^2$ area.

# ADJUSTMENT FACTORS

The Biological Weapon Effects Graphs assume specific field conditions, and require adjusting to the conditions of the target. Adjustments are made as the product of multipliers.

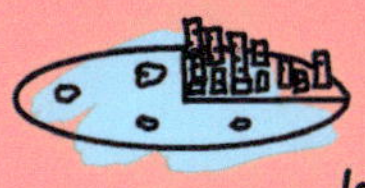

**Mixed Terrains**

When a target is composed of multiple terrain types, the overall casualty rate is a weighted average by % terrain.

(e.g., 25% Urban x 30% Casualties + 75% Open x 50% Casualties = 45%)

**Terrain & Weather**

Adjust the Source Strength (HU/bomblet) by multiplying with the factors for terrain and weather conditions other than those in the Weapon Effects Graphs.

(e.g., Hilly: $8 \times 10^8$ HU/bomblet x 0.5 = $4 \times 10^8$ HU/bomblet)

**Enemy Response**

If the enemy uses protective masks, vaccines, or medical treatment, adjust the overall casualty rate by the appropriate multiplier.

(e.g., Masking: 50% Casualties x 0.1 = 5% Casualties)

# PROBABILITY OF LEAST

The principle of POL is used in biological weapon effects estimates. By benchmarking effects for a target at rest, biological aerosols in the 1 to 5 µm diameter range and one-hour of aging, planning strength after four half-lifes, and a bomblet density for CD90 coverage, there is a high degree of confidence that casualty rates will be "at least" at or above estimates. Higher casualty rates from greater activity, larger biological aerosols received within minutes, and higher strengths are treated as a bonus.

Likewise, casualty rate estimates are made to the operational requirements of 30%, 50%, 70% or 90% and not directly to the figures indicated by the graphical solution. For example, if the graphical solution indicates a 40% casualty rate, then the estimate used in planning is 30%.

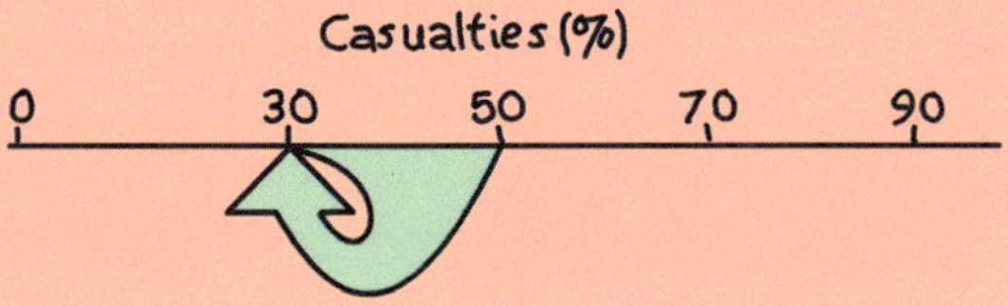


TR and OZ are not approved for operational use until the human dose-response has been confirmed.

The human median infective doses of SR, LM, and DK have been confirmed by use of WHITECOAT volunteers.

# WEAPON EFFECTS GRAPH

| TERRAIN COMPLEX | ATMOSPHERIC STABILITY | WIND SPEED | AEROBIOLOGIAL DECAY RATE |
|---|---|---|---|
| **OPEN** | **F** | **<6** m/s | **0-5** %/min |

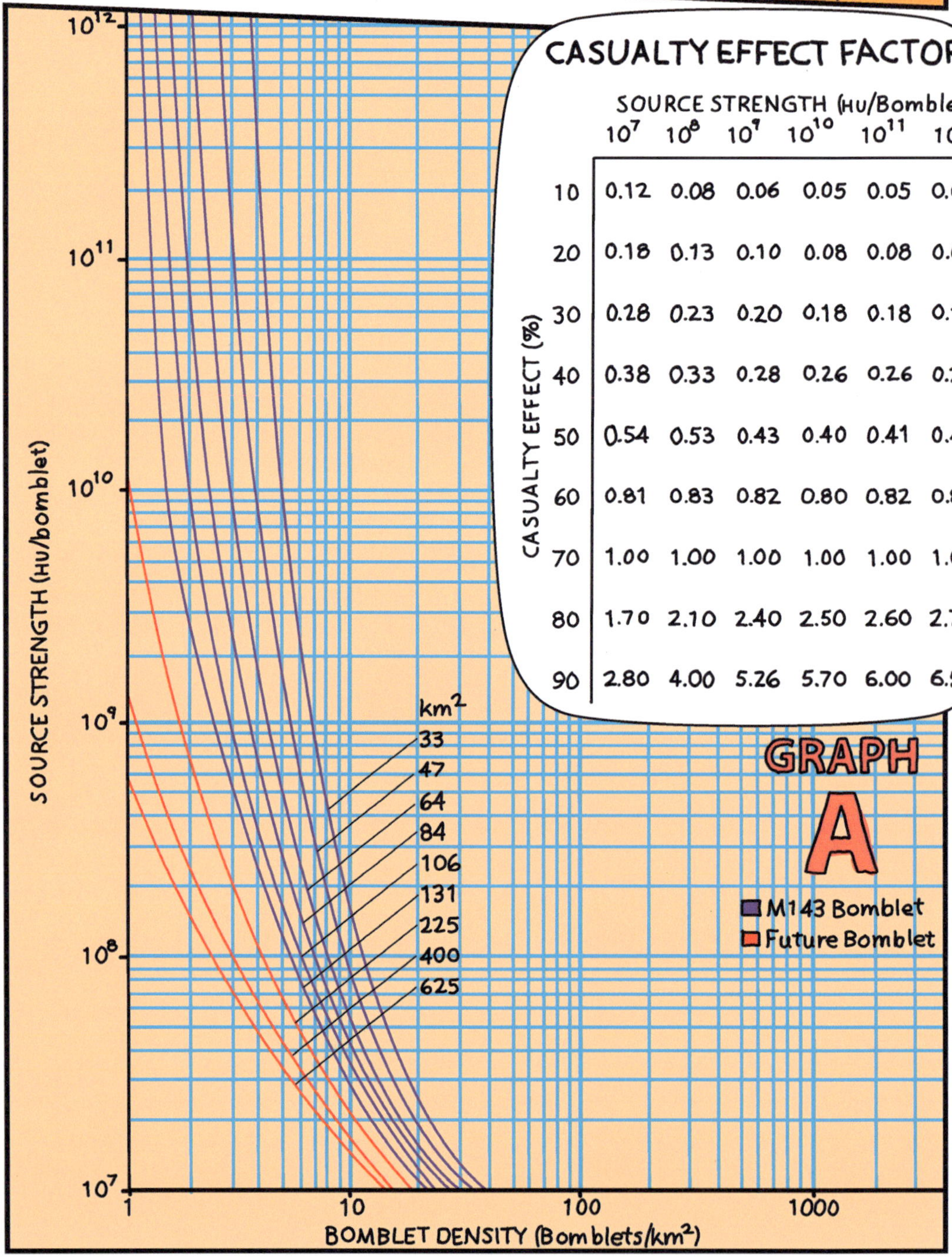


## CASUALTY EFFECT FACTOR

| CASUALTY EFFECT (%) | SOURCE STRENGTH (HU/Bomblet) $10^7$ | $10^8$ | $10^9$ | $10^{10}$ | $10^{11}$ | $10^{12}$ |
|---|---|---|---|---|---|---|
| 10 | 0.12 | 0.08 | 0.06 | 0.05 | 0.05 | 0.05 |
| 20 | 0.18 | 0.13 | 0.10 | 0.08 | 0.08 | 0.08 |
| 30 | 0.28 | 0.23 | 0.20 | 0.18 | 0.18 | 0.18 |
| 40 | 0.38 | 0.33 | 0.28 | 0.26 | 0.26 | 0.26 |
| 50 | 0.54 | 0.53 | 0.43 | 0.40 | 0.41 | 0.42 |
| 60 | 0.81 | 0.83 | 0.82 | 0.80 | 0.82 | 0.84 |
| 70 | 1.00 | 1.00 | 1.00 | 1.00 | 1.00 | 1.00 |
| 80 | 1.70 | 2.10 | 2.40 | 2.50 | 2.60 | 2.70 |
| 90 | 2.80 | 4.00 | 5.26 | 5.70 | 6.00 | 6.50 |

# WEAPON EFFECTS GRAPH

| TERRAIN COMPLEX | ATMOSPHERIC STABILITY | WIND SPEED | AEROBIOLOGIAL DECAY RATE |
|---|---|---|---|
| **OPEN** | **D** | **>0** m/s | **0-5** %/min |

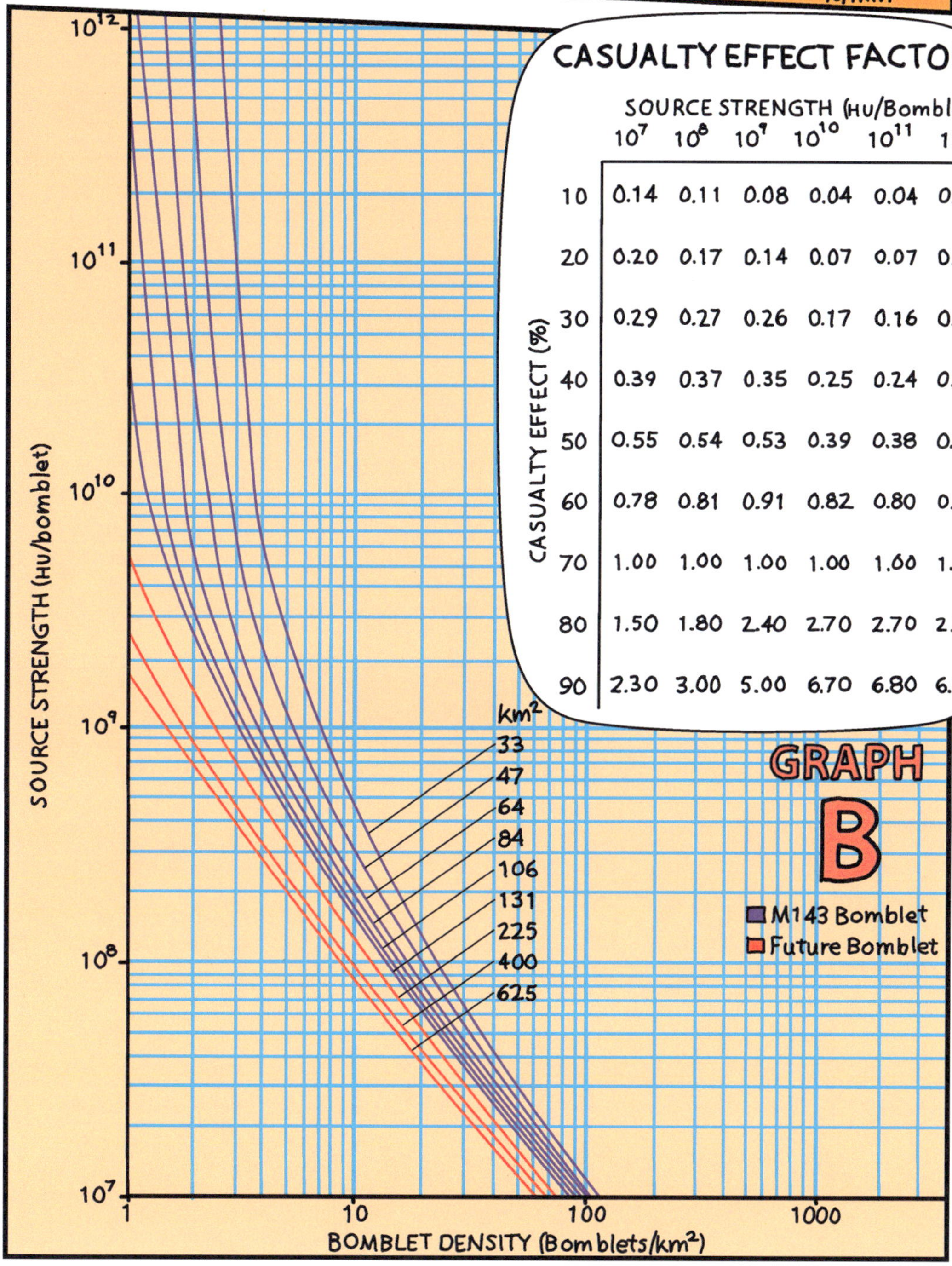


## CASUALTY EFFECT FACTOR

| CASUALTY EFFECT (%) | SOURCE STRENGTH (HU/Bomblet) $10^{7}$ | $10^{8}$ | $10^{9}$ | $10^{10}$ | $10^{11}$ | $10^{12}$ |
|---|---|---|---|---|---|---|
| 10 | 0.14 | 0.11 | 0.08 | 0.04 | 0.04 | 0.04 |
| 20 | 0.20 | 0.17 | 0.14 | 0.07 | 0.07 | 0.07 |
| 30 | 0.29 | 0.27 | 0.26 | 0.17 | 0.16 | 0.16 |
| 40 | 0.39 | 0.37 | 0.35 | 0.25 | 0.24 | 0.24 |
| 50 | 0.55 | 0.54 | 0.53 | 0.39 | 0.38 | 0.39 |
| 60 | 0.78 | 0.81 | 0.91 | 0.82 | 0.80 | 0.80 |
| 70 | 1.00 | 1.00 | 1.00 | 1.00 | 1.60 | 1.00 |
| 80 | 1.50 | 1.80 | 2.40 | 2.70 | 2.70 | 2.70 |
| 90 | 2.30 | 3.00 | 5.00 | 6.70 | 6.80 | 6.70 |

# WEAPON EFFECTS GRAPH

| TERRAIN COMPLEX | ATMOSPHERIC STABILITY | WIND SPEED | AEROBIOLOGIAL DECAY RATE |
|---|---|---|---|
| **URBAN** | **F** | **<6** m/s | **0 - 5** %/min |

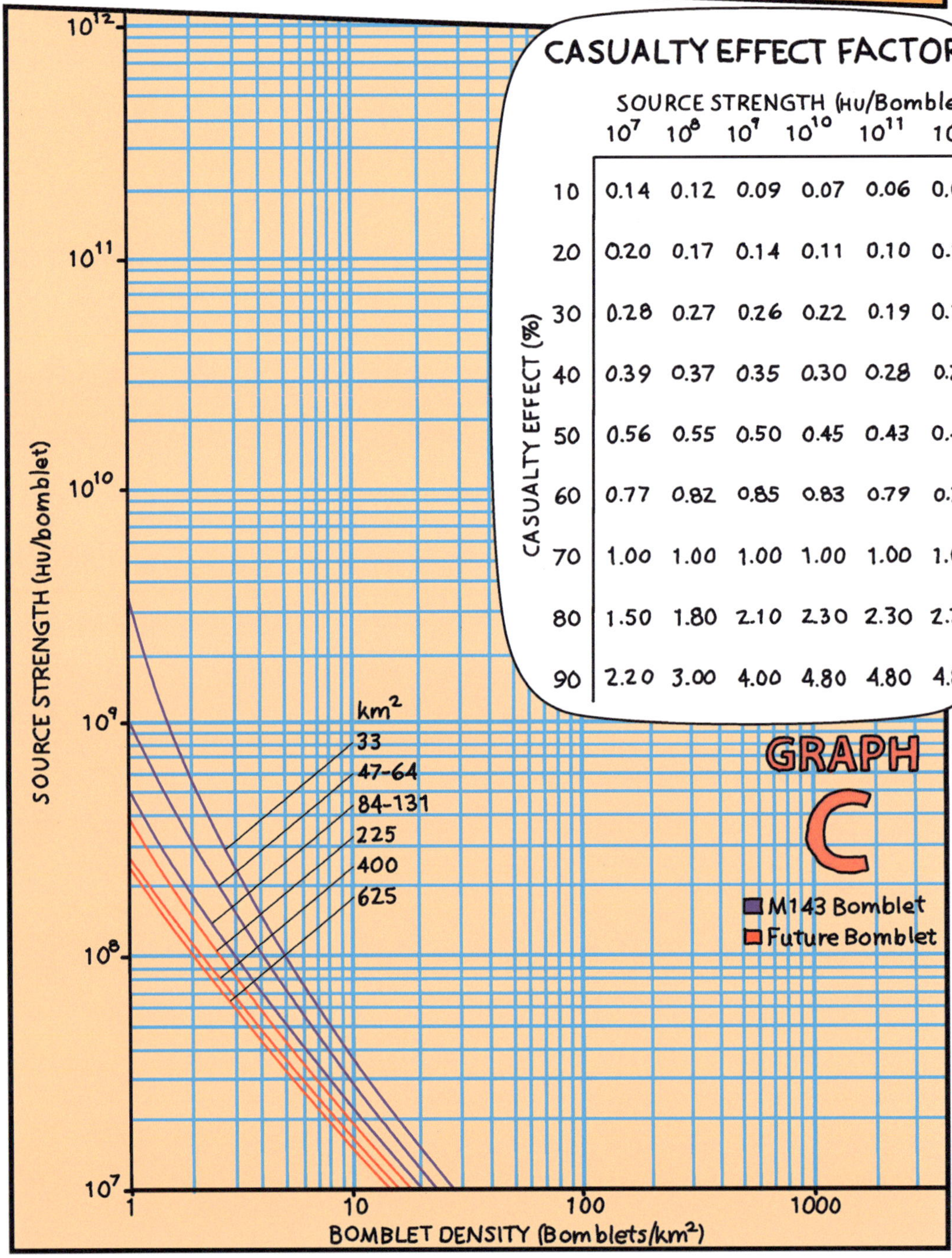


## CASUALTY EFFECT FACTOR

| CASUALTY EFFECT (%) | SOURCE STRENGTH (HU/Bomblet) $10^7$ | $10^8$ | $10^9$ | $10^{10}$ | $10^{11}$ | $10^{12}$ |
|---|---|---|---|---|---|---|
| 10 | 0.14 | 0.12 | 0.09 | 0.07 | 0.06 | 0.06 |
| 20 | 0.20 | 0.17 | 0.14 | 0.11 | 0.10 | 0.10 |
| 30 | 0.28 | 0.27 | 0.26 | 0.22 | 0.19 | 0.19 |
| 40 | 0.39 | 0.37 | 0.35 | 0.30 | 0.28 | 0.28 |
| 50 | 0.56 | 0.55 | 0.50 | 0.45 | 0.43 | 0.43 |
| 60 | 0.77 | 0.82 | 0.85 | 0.83 | 0.79 | 0.79 |
| 70 | 1.00 | 1.00 | 1.00 | 1.00 | 1.00 | 1.00 |
| 80 | 1.50 | 1.80 | 2.10 | 2.30 | 2.30 | 2.30 |
| 90 | 2.20 | 3.00 | 4.00 | 4.80 | 4.80 | 4.80 |

# WEAPON EFFECTS GRAPH

| TERRAIN COMPLEX | ATMOSPHERIC STABILITY | WIND SPEED | AEROBIOLOGIAL DECAY RATE |
|---|---|---|---|
| **URBAN** | **D** | **>0** m/s | **0-5** %/min |

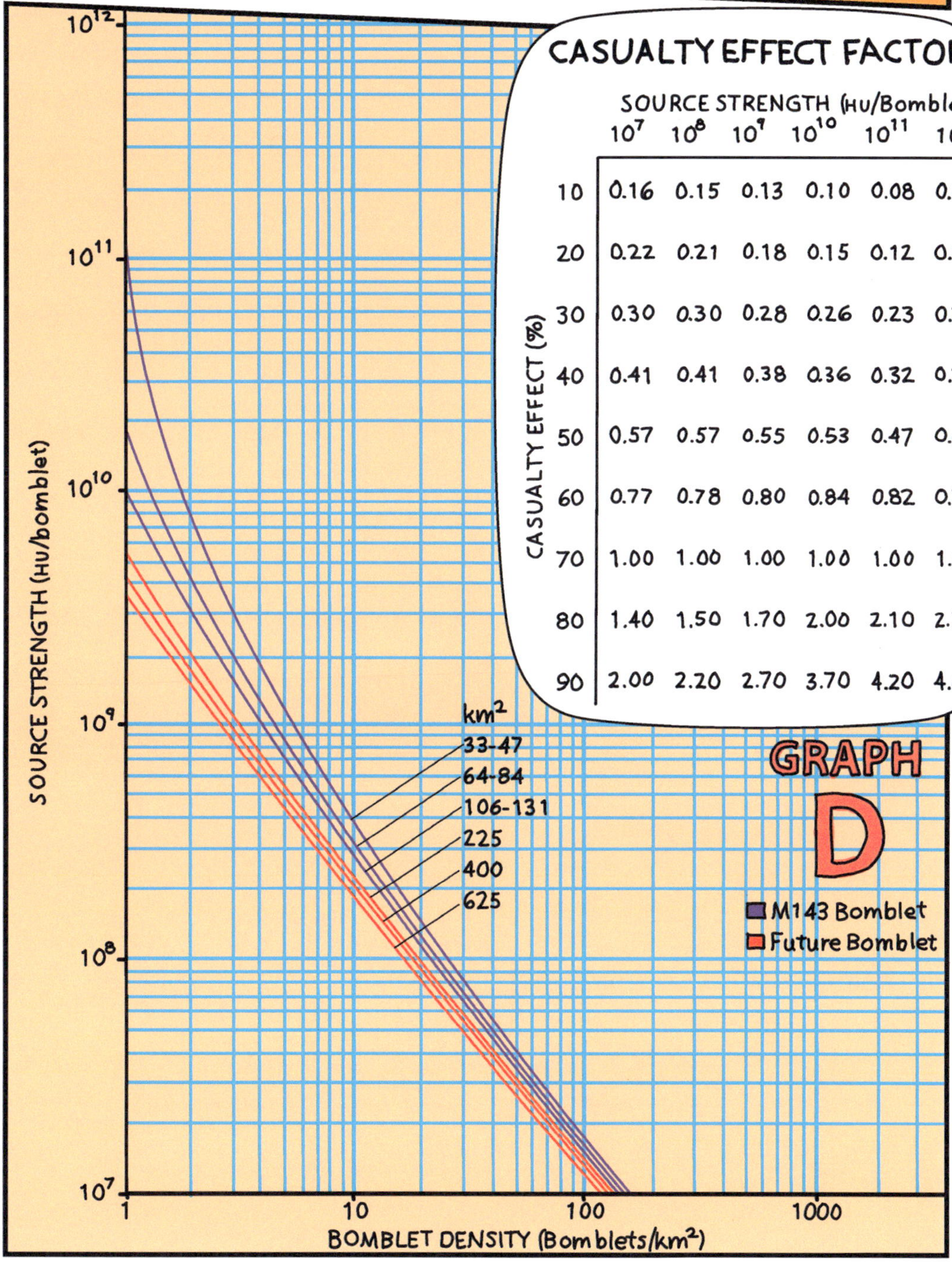


## CASUALTY EFFECT FACTOR

| CASUALTY EFFECT (%) | SOURCE STRENGTH (HU/Bomblet) $10^{7}$ | $10^{8}$ | $10^{9}$ | $10^{10}$ | $10^{11}$ | $10^{12}$ |
|---|---|---|---|---|---|---|
| 10 | 0.16 | 0.15 | 0.13 | 0.10 | 0.08 | 0.06 |
| 20 | 0.22 | 0.21 | 0.18 | 0.15 | 0.12 | 0.10 |
| 30 | 0.30 | 0.30 | 0.28 | 0.26 | 0.23 | 0.20 |
| 40 | 0.41 | 0.41 | 0.38 | 0.36 | 0.32 | 0.28 |
| 50 | 0.57 | 0.57 | 0.55 | 0.53 | 0.47 | 0.43 |
| 60 | 0.77 | 0.78 | 0.80 | 0.84 | 0.82 | 0.79 |
| 70 | 1.00 | 1.00 | 1.00 | 1.00 | 1.00 | 1.00 |
| 80 | 1.40 | 1.50 | 1.70 | 2.00 | 2.10 | 2.20 |
| 90 | 2.00 | 2.20 | 2.70 | 3.70 | 4.20 | 4.60 |

# WEAPON EFFECTS GRAPH

| TERRAIN COMPLEX | ATMOSPHERIC STABILITY | WIND SPEED | AEROBIOLOGIAL DECAY RATE |
|---|---|---|---|
| **OPEN/URBAN** | **G-F** | **<4** m/s | **20** %/min |

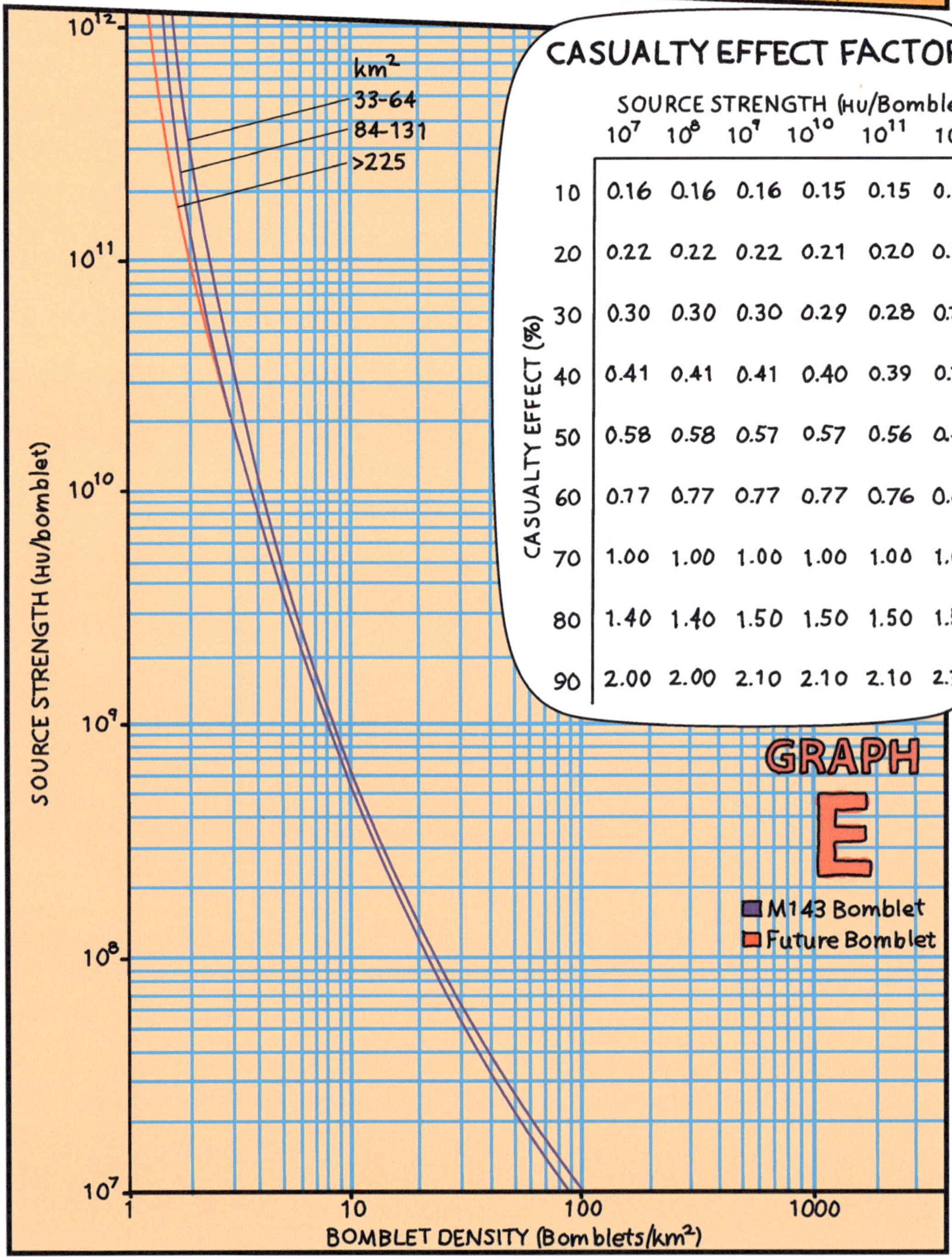


## CASUALTY EFFECT FACTOR

| CASUALTY EFFECT (%) | SOURCE STRENGTH (HU/Bomblet) $10^7$ | $10^8$ | $10^9$ | $10^{10}$ | $10^{11}$ | $10^{12}$ |
|---|---|---|---|---|---|---|
| 10 | 0.16 | 0.16 | 0.16 | 0.15 | 0.15 | 0.09 |
| 20 | 0.22 | 0.22 | 0.22 | 0.21 | 0.20 | 0.14 |
| 30 | 0.30 | 0.30 | 0.30 | 0.29 | 0.28 | 0.20 |
| 40 | 0.41 | 0.41 | 0.41 | 0.40 | 0.39 | 0.30 |
| 50 | 0.58 | 0.58 | 0.57 | 0.57 | 0.56 | 0.46 |
| 60 | 0.77 | 0.77 | 0.77 | 0.77 | 0.76 | 0.68 |
| 70 | 1.00 | 1.00 | 1.00 | 1.00 | 1.00 | 1.00 |
| 80 | 1.40 | 1.40 | 1.50 | 1.50 | 1.50 | 1.50 |
| 90 | 2.00 | 2.00 | 2.10 | 2.10 | 2.10 | 2.30 |

# WEAPON EFFECTS GRAPH

| TERRAIN COMPLEX | ATMOSPHERIC STABILITY | WIND SPEED | AEROBIOLOGIAL DECAY RATE |
|---|---|---|---|
| OPEN/URBAN | F-D | >4 m/s | 20 %/min |

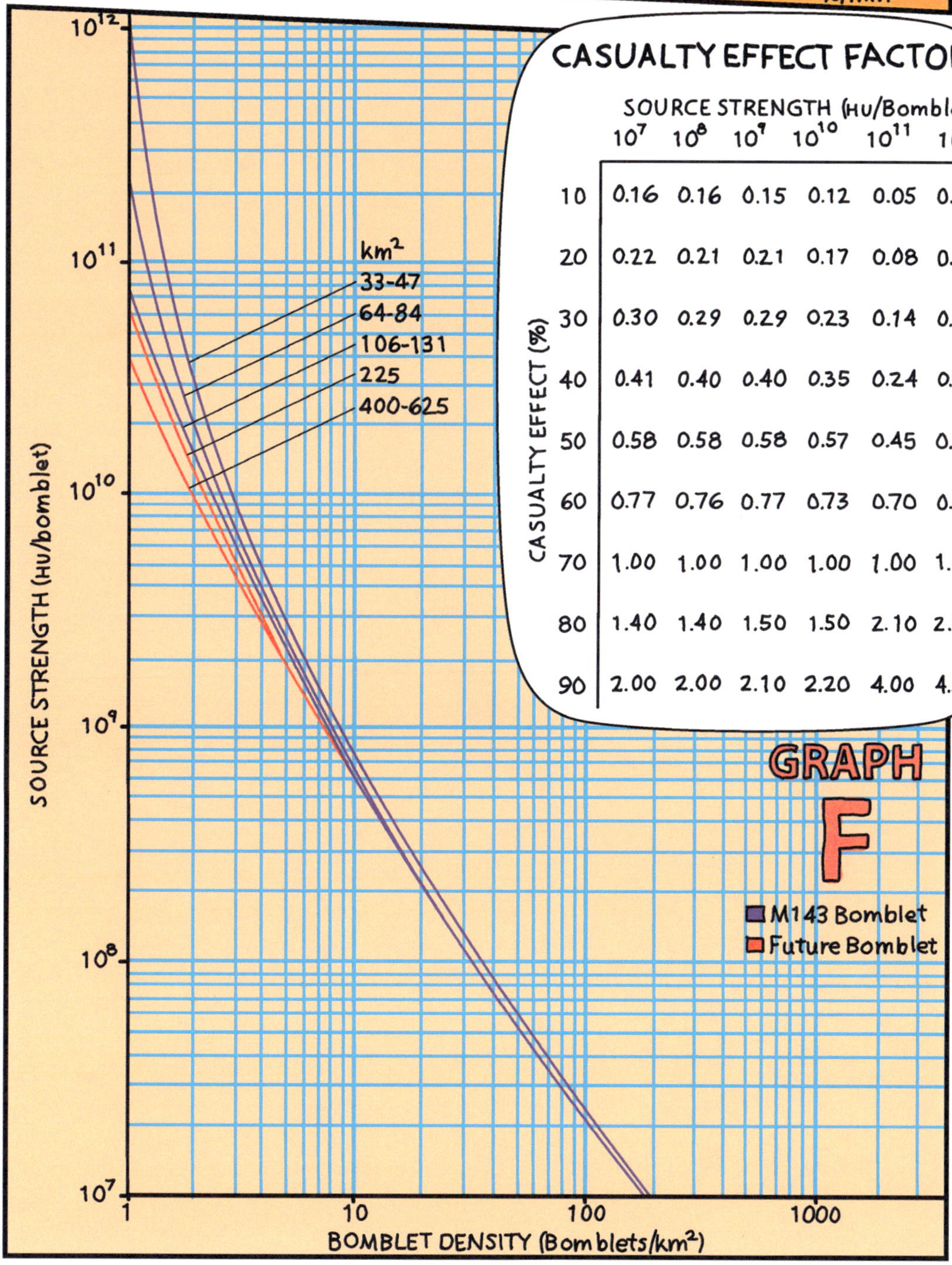


## CASUALTY EFFECT FACTOR

| CASUALTY EFFECT (%) | SOURCE STRENGTH (HU/Bomblet) $10^7$ | $10^8$ | $10^9$ | $10^{10}$ | $10^{11}$ | $10^{12}$ |
|---|---|---|---|---|---|---|
| 10 | 0.16 | 0.16 | 0.15 | 0.12 | 0.05 | 0.04 |
| 20 | 0.22 | 0.21 | 0.21 | 0.17 | 0.08 | 0.07 |
| 30 | 0.30 | 0.29 | 0.29 | 0.23 | 0.14 | 0.12 |
| 40 | 0.41 | 0.40 | 0.40 | 0.35 | 0.24 | 0.21 |
| 50 | 0.58 | 0.58 | 0.58 | 0.57 | 0.45 | 0.43 |
| 60 | 0.77 | 0.76 | 0.77 | 0.73 | 0.70 | 0.69 |
| 70 | 1.00 | 1.00 | 1.00 | 1.00 | 1.00 | 1.00 |
| 80 | 1.40 | 1.40 | 1.50 | 1.50 | 2.10 | 2.20 |
| 90 | 2.00 | 2.00 | 2.10 | 2.20 | 4.00 | 4.50 |

# MEDICAL SUPPORT

## TREATMENT

Tularemia casualties require hospitalization for treatment. Doxycycline is given orally in 500mg doses four times a day for a minimum of two weeks. A shorter course of treatment will result in a relapse of infection. Treatment will reduce lethality to less than 1% and shorten the course of infection to about a month.

Quarantine and isolation are not required

## PROPHYLAXIS

Oral Doxycycline may be given to prevent infection up to 24 hours after an attack with Tularemia. It must be given in 500 mg doses twice a day for a month or 1,000 mg doses twice a day for 15 days. Tularemia infection will resume if prophylaxis is discontinued or reduced.

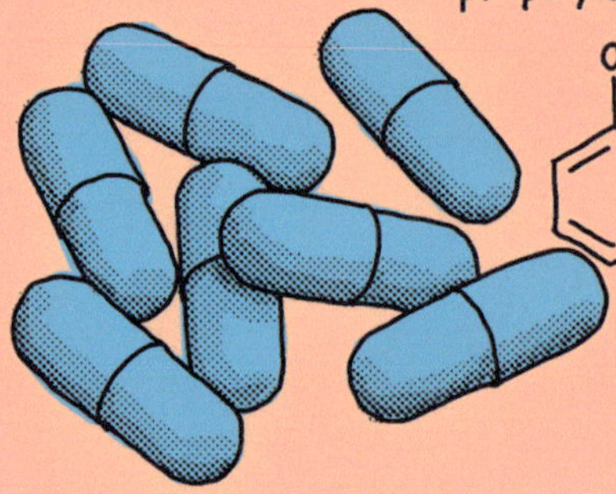

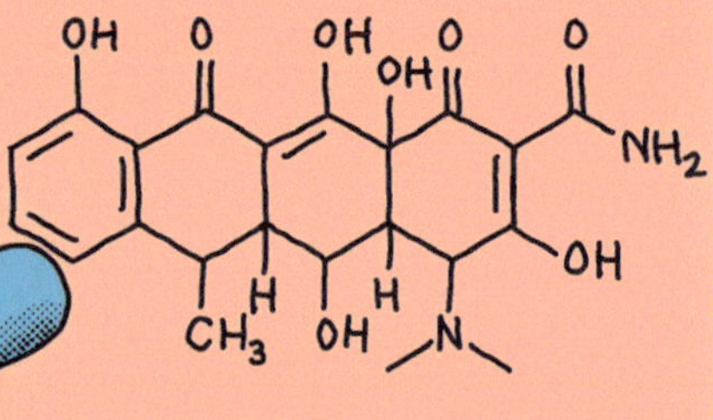

## VACCINATION

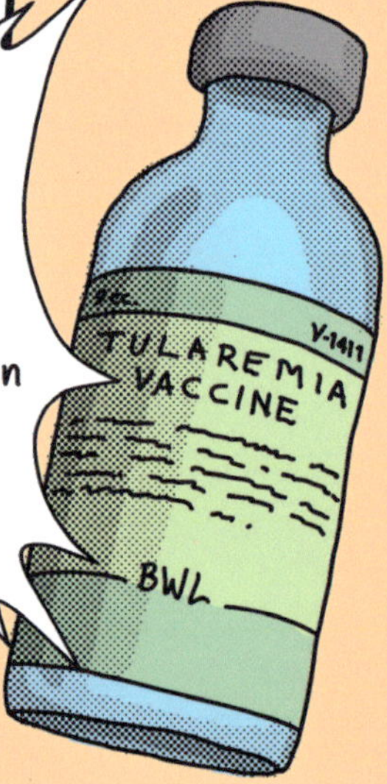


The BWL has developed a Live Vaccine Strain (LVS) of Tularemia (MO) that is 80% effective in preventing infection.

Troops are vaccinated enmasse through aerogenic administration of MO two weeks before biological operations. Vaccine is protective for up to 2 years.

Use of prophylaxis or vaccination increases freedom of action on the biological battlefield.

The median effective dose is 600 org/person. Up to $10^4$ org/person can be received without symptoms, and $10^8$ org/person will produce minor illness 3 to 4 days after administration.

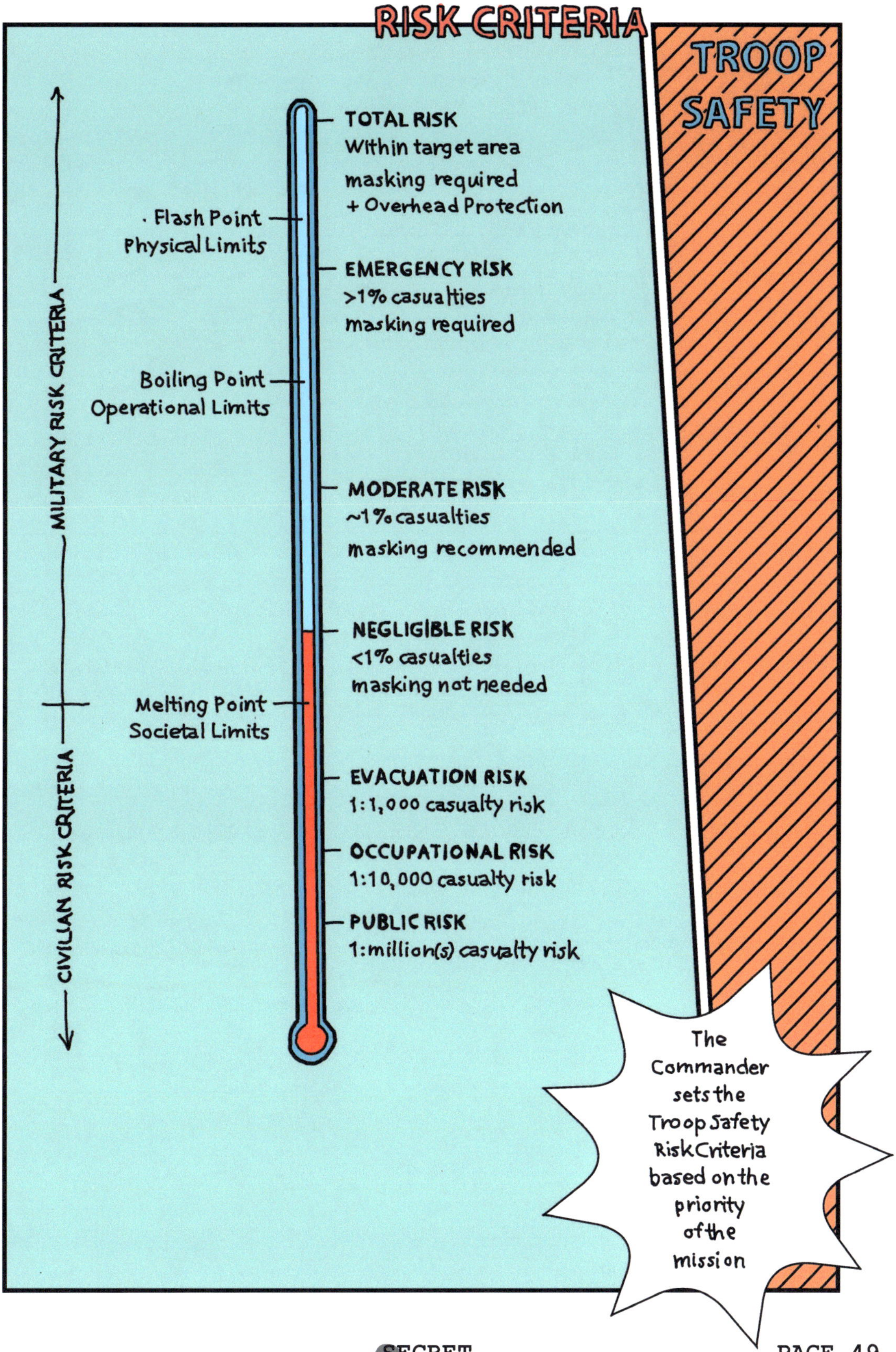
RISK CRITERIA
TROOP SAFETY
MILITARY RISK CRITERIA
CIVILIAN RISK CRITERIA
TOTAL RISK
Within target area
masking required
+ Overhead Protection
Flash Point
Physical Limits
EMERGENCY RISK
>1% casualties
masking required
Boiling Point
Operational Limits
MODERATE RISK
~1% casualties
masking recommended
NEGLIGIBLE RISK
<1% casualties
masking not needed
Melting Point
Societal Limits
EVACUATION RISK
1:1,000 casualty risk
OCCUPATIONAL RISK
1:10,000 casualty risk
PUBLIC RISK
1:million(s) casualty risk
The Commander sets the Troop Safety Risk Criteria based on the priority of the mission

# MINIMUM SAFE DISTANCE

## IMPACT AREA

Due to delivery errors, 2.5x of the CEP is added to the RD. For the Sergeant this is BUFFER of 760 meters, or 0.8 km

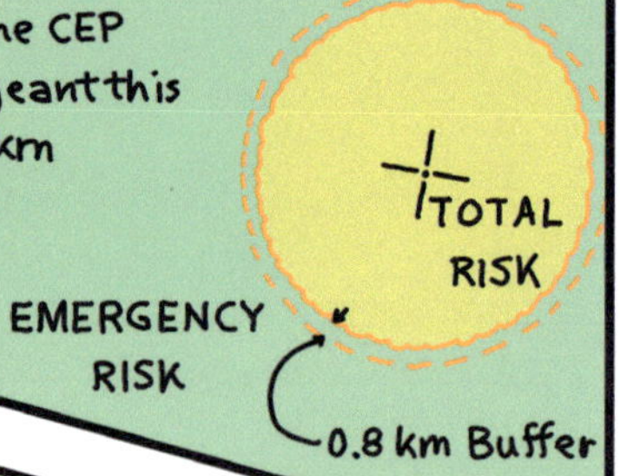


## DOWNWIND

The DOWNWIND DISTANCE is estimated using an equivalent line source at half the RD that is twice the length of the RD.

Using nomograms, the downwind distance is based on 1% casualty effect for 274 ml viable biological aerosol released along equivalent line source.

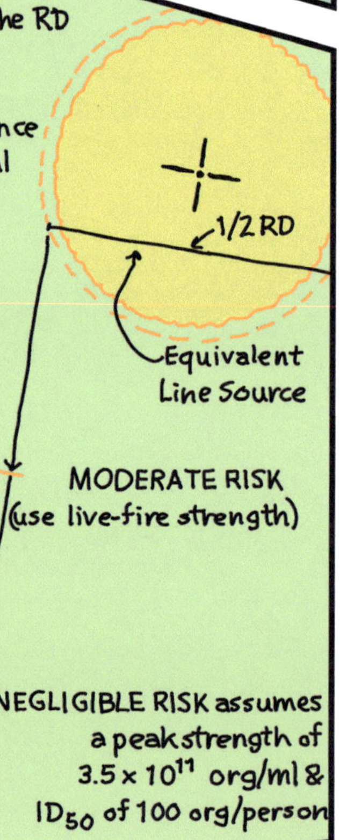


NEGLIGIBLE RISK

| HOB (feet) | Equivalent Line Length (m) | Source Strength (HU/m) |
|---|---|---|
| 25,000 | 6,400 | $3.0 \times 10^8$ |
| 30,000 | 7,800 | $2.5 \times 10^8$ |
| 35,000 | 9,000 | $2.1 \times 10^8$ |
| 40,000 | 10,400 | $1.8 \times 10^8$ |
| 45,000 | 11,600 | $1.7 \times 10^8$ |
| 50,000 | 13,000 | $1.5 \times 10^8$ |

Per M210

Add 0.8 km Buffer to downwind distances.

## SPREAD

| Angle From Wind Direction (Degrees) | Downwind Distance (%) |
|---|---|
| 0.0 | 100 |
| 45.0 | 90 |
| 67.5 | 67 |
| 90.0 | 35 |

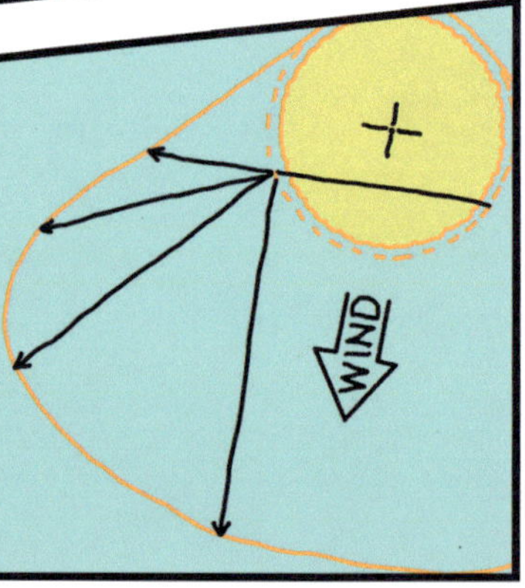

## EXAMPLE

Create MAP OVERLAY as on next page.

NEGLIGIBLE RISK DOWNWIND DISTANCE–

**STEP 1** Source Strength (6.5 km RD)

The equivalent line length is 13,000 m, indicating $1.5 \times 10^8$ HU/m.

**STEP 2** Pivot Point (Neutral Atmospheric Stability)

Using the Ground Level Line Source Nomogram for OPEN Terrain, draw a line from $1.5 \times 10^8$ HU/m to Atmospheric Stability to indicate a point on Reference Line.

**STEP 3** Nomogram Entry Number (1% Casualty Effect)

Draw a line from Pivot Point to Casualty Effect to indicate a Nomogram Entry Number of 15.

**STEP 4** Aerobiological Decay Factor (5%/min)

Dividing the Aerobiological Decay Rate by the windspeed indicates a factor of 0.5.

**STEP 5** Downwind Distance (Neutral)

Using the B portion of nomogram, draw a line from Nomogram Entry Number 15 to Aerobiological Decay Factor of 0.5 to indicate 20 km on Neutral Line (add 0.8 km Buffer).

MODERATE RISK DOWNWIND DISTANCE – repeat steps, except using a live-fire strength of $9.4 \times 10^6$ HU/m.

# MAP OVERLAY

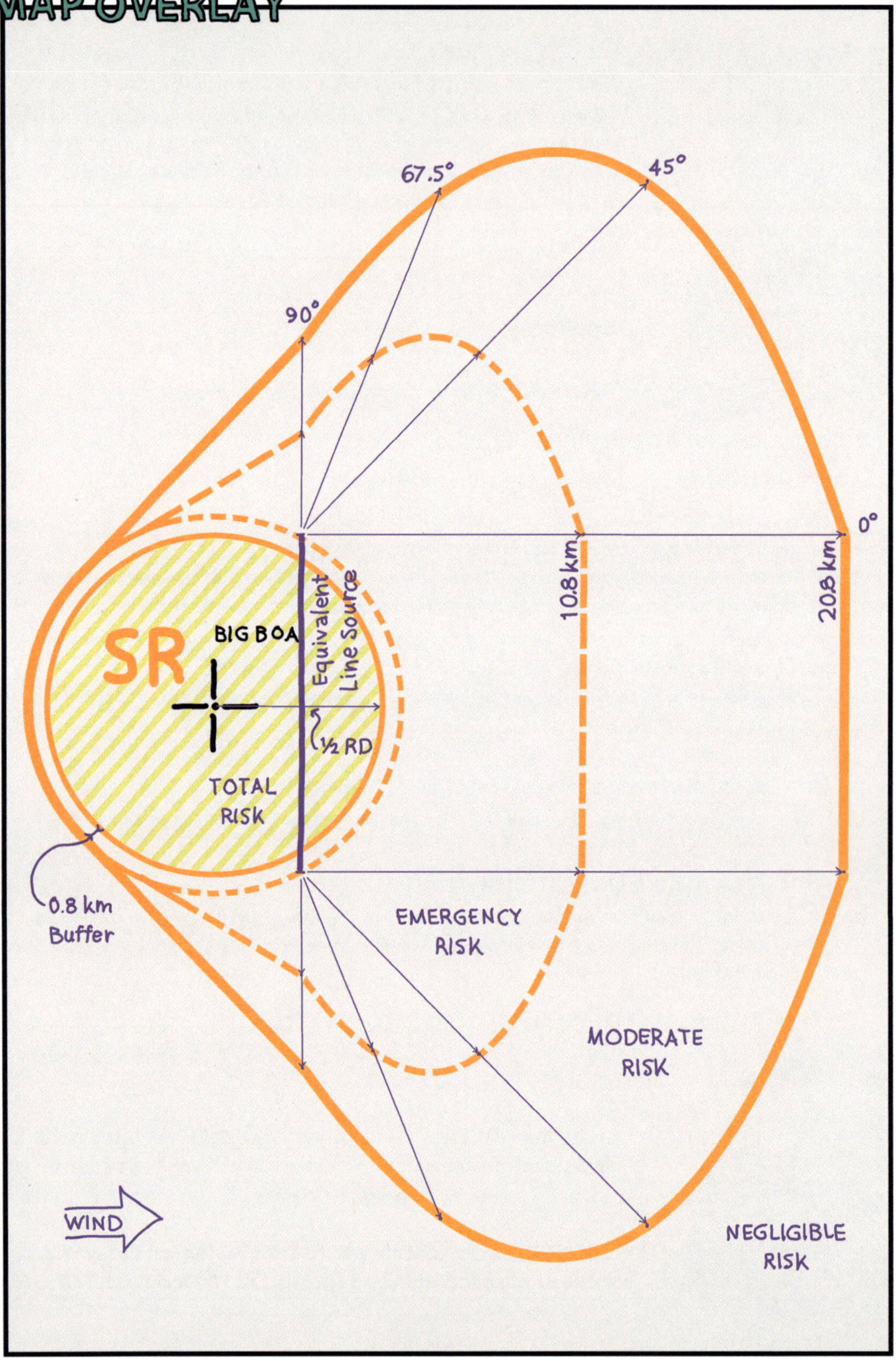
67.5°
45°
90°
0°
10.8 km
20.8 km
Equivalent Line Source
BIG BOA
SR
½ RD
TOTAL RISK
0.8 km Buffer
EMERGENCY RISK
MODERATE RISK
WIND
NEGLIGIBLE RISK

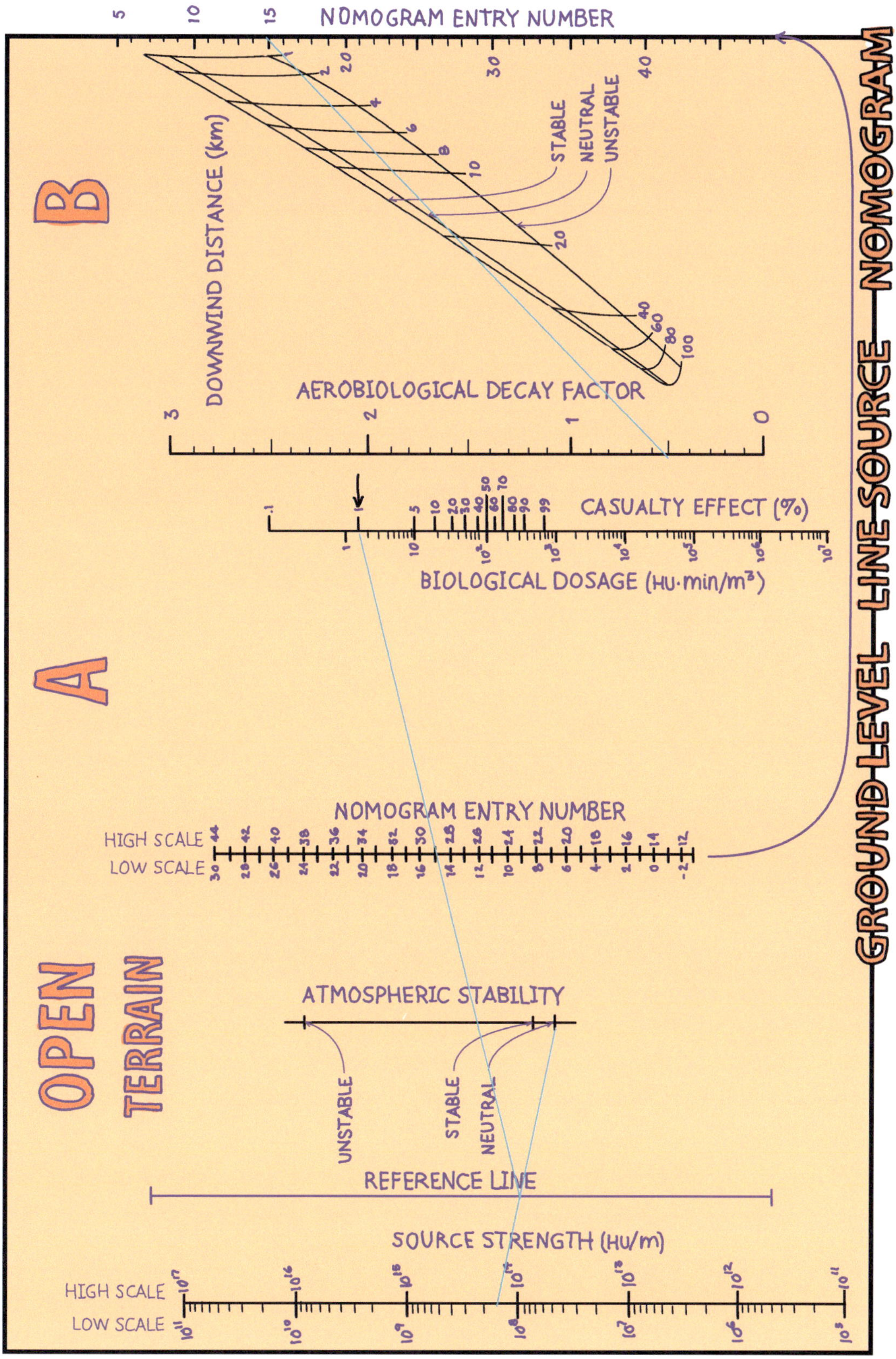
GROUND LEVEL LINE SOURCE NOMOGRAM
OPEN TERRAIN
A
B
NOMOGRAM ENTRY NUMBER
DOWNWIND DISTANCE (km)
STABLE
NEUTRAL
UNSTABLE
AEROBIOLOGICAL DECAY FACTOR
CASUALTY EFFECT (%)
BIOLOGICAL DOSAGE (HU·min/m³)
NOMOGRAM ENTRY NUMBER
HIGH SCALE
LOW SCALE
ATMOSPHERIC STABILITY
UNSTABLE
STABLE
NEUTRAL
REFERENCE LINE
SOURCE STRENGTH (HU/m)
HIGH SCALE
LOW SCALE

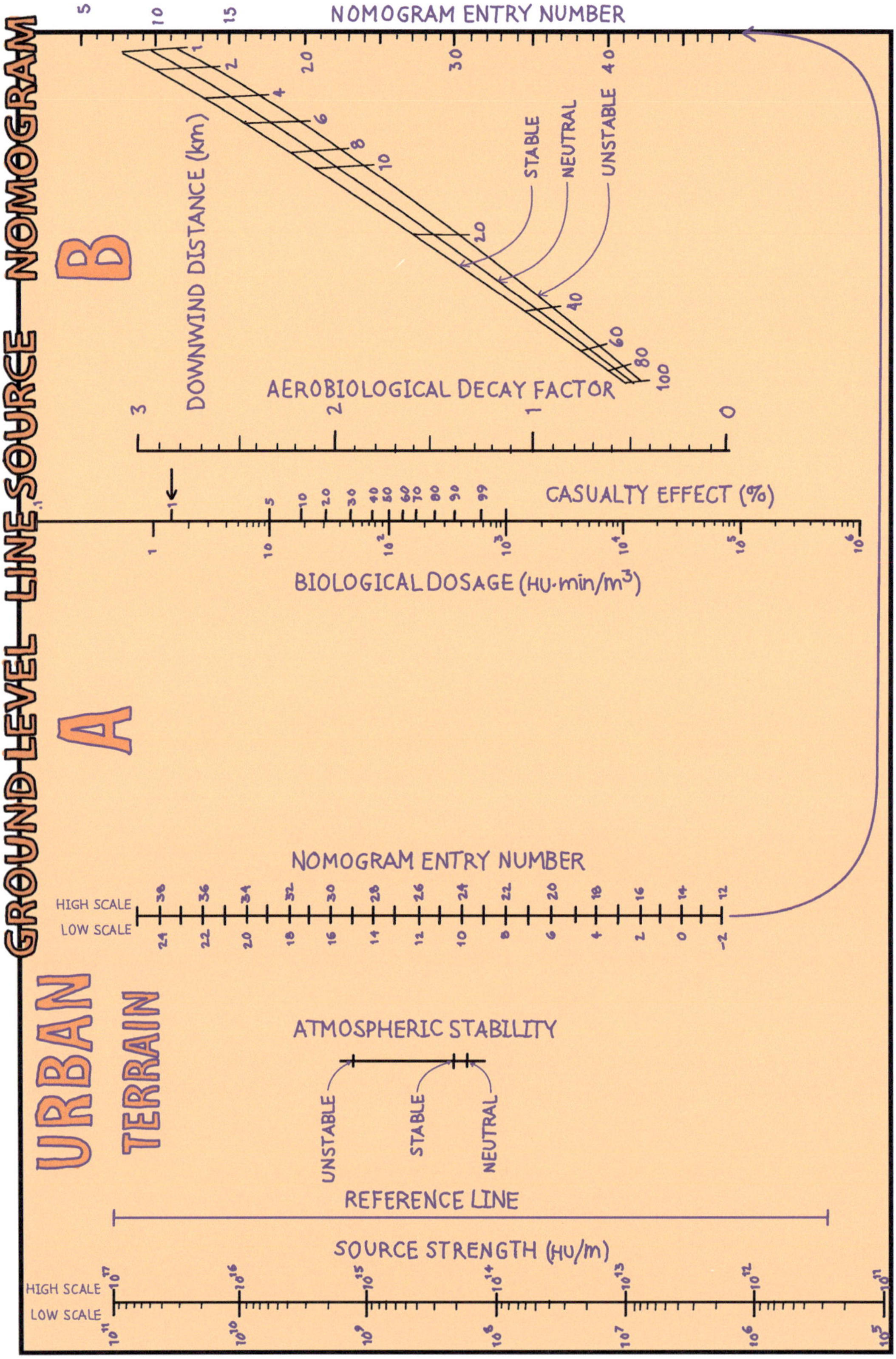
GROUND LEVEL LINE SOURCE NOMOGRAM
URBAN TERRAIN
A
B
NOMOGRAM ENTRY NUMBER
DOWNWIND DISTANCE (km)
STABLE
NEUTRAL
UNSTABLE
AEROBIOLOGICAL DECAY FACTOR
CASUALTY EFFECT (%)
BIOLOGICAL DOSAGE (HU·min/m3)
NOMOGRAM ENTRY NUMBER
HIGH SCALE
LOW SCALE
ATMOSPHERIC STABILITY
UNSTABLE
STABLE
NEUTRAL
REFERENCE LINE
SOURCE STRENGTH (HU/m)
HIGH SCALE
LOW SCALE

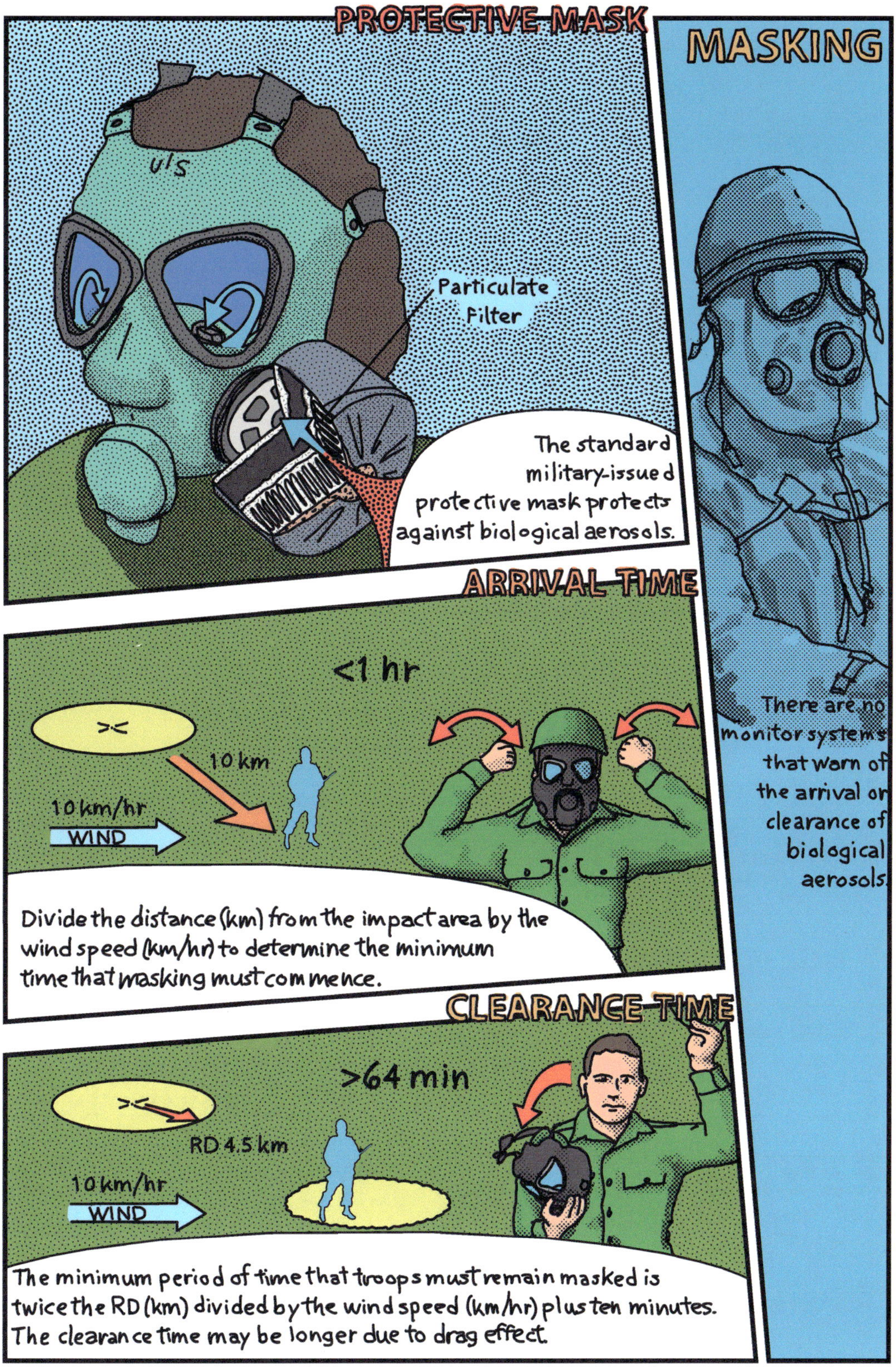
PROTECTIVE MASK
U/S
Particulate Filter
The standard military-issued protective mask protects against biological aerosols.
MASKING
There are no monitor systems that warn of the arrival or clearance of biological aerosols.
ARRIVAL TIME
<1 hr
10 km
10 km/hr
WIND
Divide the distance (km) from the impact area by the wind speed (km/hr) to determine the minimum time that masking must commence.
CLEARANCE TIME
>64 min
RD 4.5 km
10 km/hr
WIND
The minimum period of time that troops must remain masked is twice the RD (km) divided by the wind speed (km/hr) plus ten minutes. The clearance time may be longer due to drag effect.

# BIOWARN

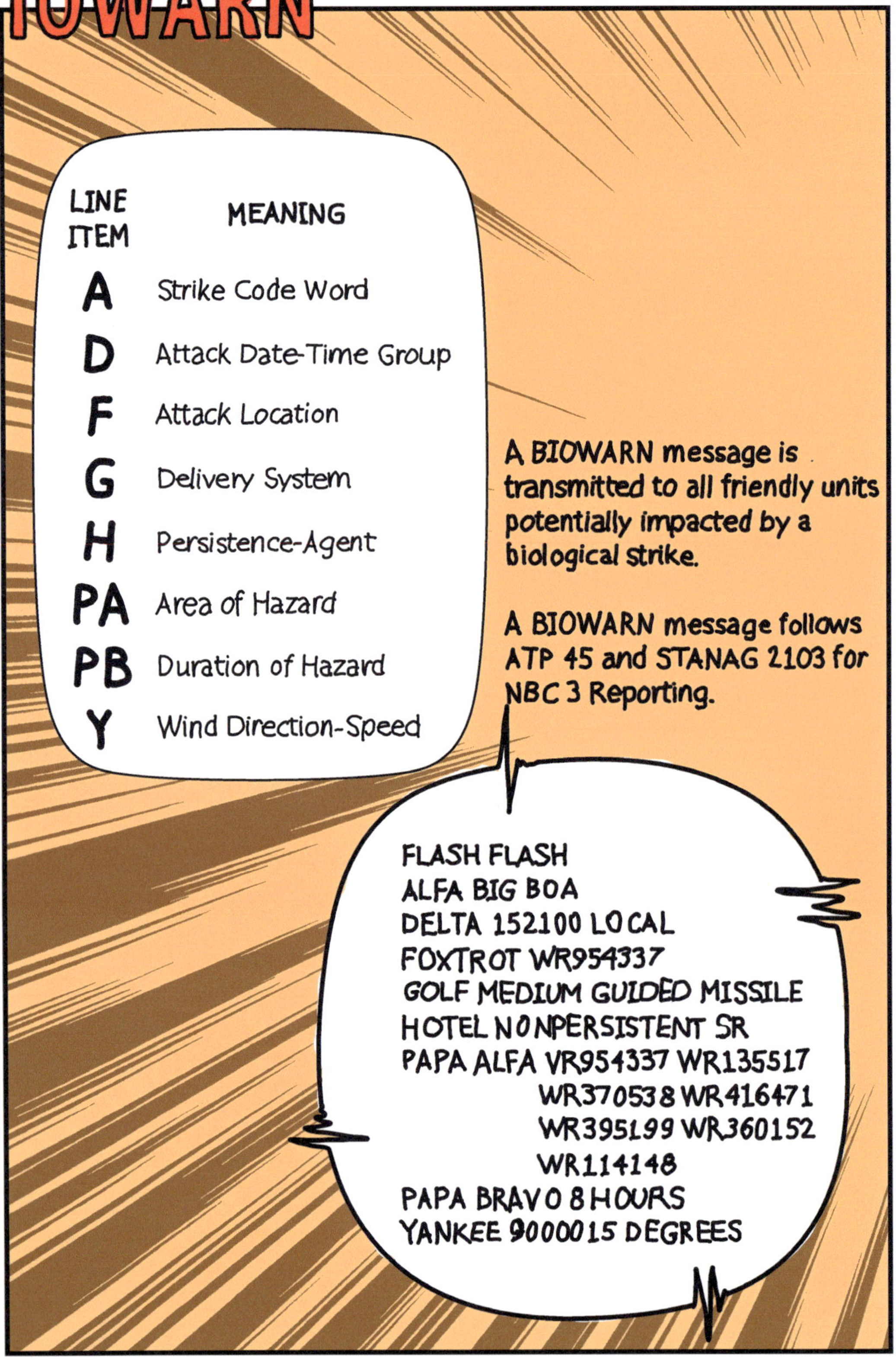
LINE ITEM
MEANING
A Strike Code Word
D Attack Date-Time Group
F Attack Location
G Delivery System
H Persistence-Agent
PA Area of Hazard
PB Duration of Hazard
Y Wind Direction-Speed
A BIOWARN message is transmitted to all friendly units potentially impacted by a biological strike.
A BIOWARN message follows ATP 45 and STANAG 2103 for NBC 3 Reporting.
FLASH FLASH
ALFA BIG BOA
DELTA 152100 LOCAL
FOXTROT WR954337
GOLF MEDIUM GUIDED MISSILE
HOTEL NONPERSISTENT SR
PAPA ALFA VR954337 WR135517
WR370538 WR416471
WR395199 WR360152
WR114148
PAPA BRAVO 8 HOURS
YANKEE 9000015 DEGREES

## TERRAIN

Open terrain is normally safe to occupy after 24 hours. Most biological aerosols are completely destroyed after 2 hours' contact with sunlight, or after 8 hours at night or overcast skies. Some biologicals, like TR and LM, may require up to 18 hours' exposure during night or overcasts.

There may be limited places that are a special hazard for several days. These are typically within the bomblet impact area where biological aerosols have settled out in dark and dusty locations.

Decontamination is not normally required after biological attacks

## FACILITIES

β-Propiolactone (BPL) is dispersed inside suspected contaminated facilities from a push cart sprayer. 1 gallon will decontaminate 25,000 cubic feet.

Humidity must be above 70%. BPL requires 2 hours' contact time at 70°F (double time for every 18°F below 70°F). After 24 hours of aeration, facility is safe to occupy.

## EQUIPMENT

Personal equipment and vehicles are decontaminated with Ethylene Oxide (ETO) from the M10 12-ounce aerosol dispenser. One M10 is used on personal equipment in a 2.5-foot by 6-foot plastic bag, and 2 M10s for a vehicle under a tarpaulin.

Do not use ETO below 60°F. Contact time is 6 to 8 hours for 70°F, and 12 hours at 60°F. Aerate items for 12 hours before use (18 hours for items in contact with skin).

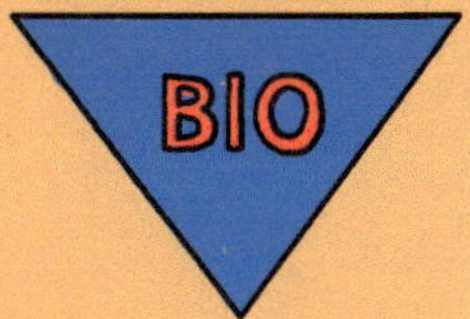


In special cases where heavy contamination is suspected Use the BIO sign to mark-off terrain, facilities, and equipment requiring decontamination.

## DECON

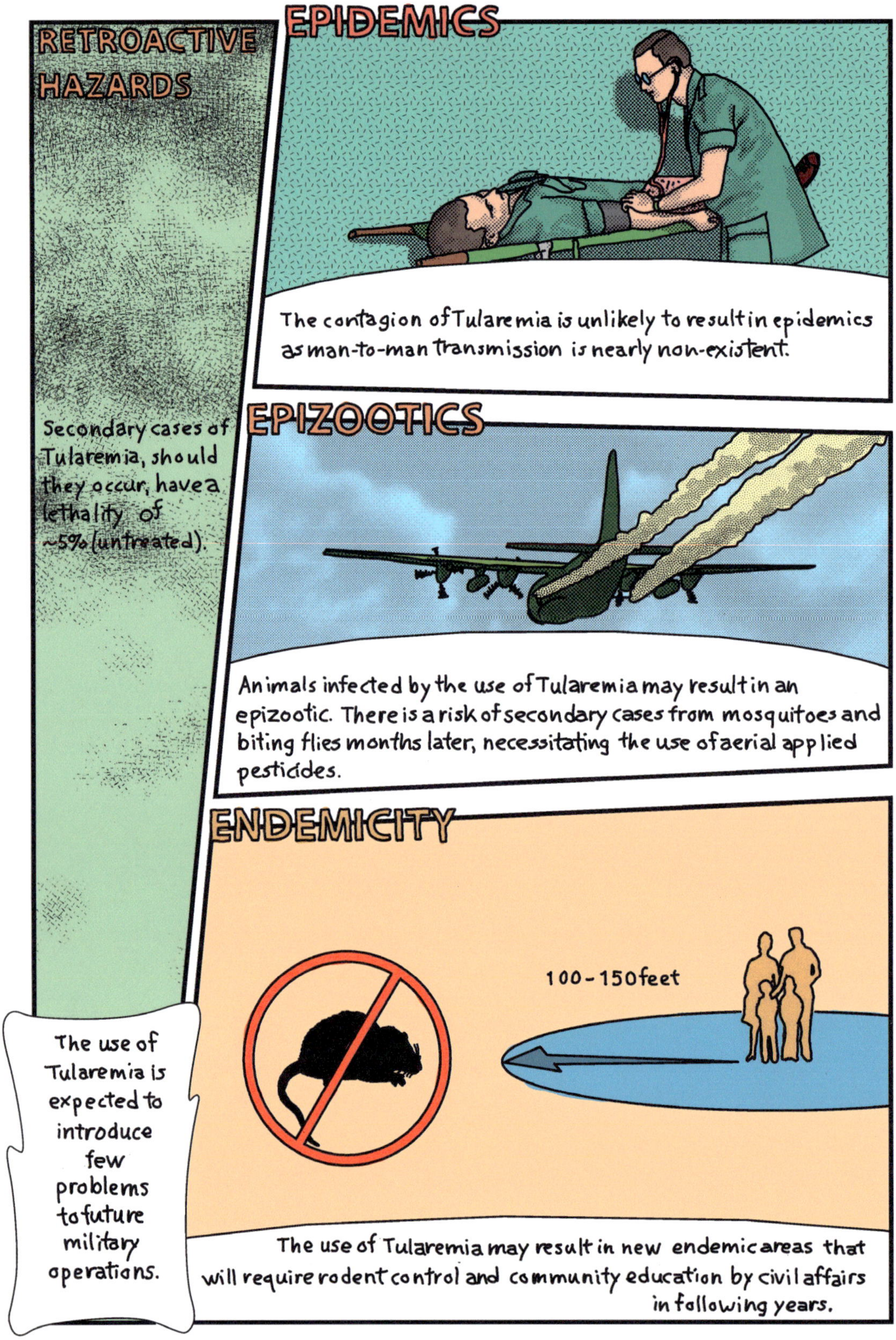

RETROACTIVE HAZARDS
EPIDEMICS
The contagion of Tularemia is unlikely to result in epidemics as man-to-man transmission is nearly non-existent.
Secondary cases of Tularemia, should they occur, have a lethality of ~5% (untreated).
EPIZOOTICS
Animals infected by the use of Tularemia may result in an epizootic. There is a risk of secondary cases from mosquitoes and biting flies months later, necessitating the use of aerial applied pesticides.
ENDEMICITY
100-150 feet
The use of Tularemia is expected to introduce few problems to future military operations.
The use of Tularemia may result in new endemic areas that will require rodent control and community education by civil affairs in following years.

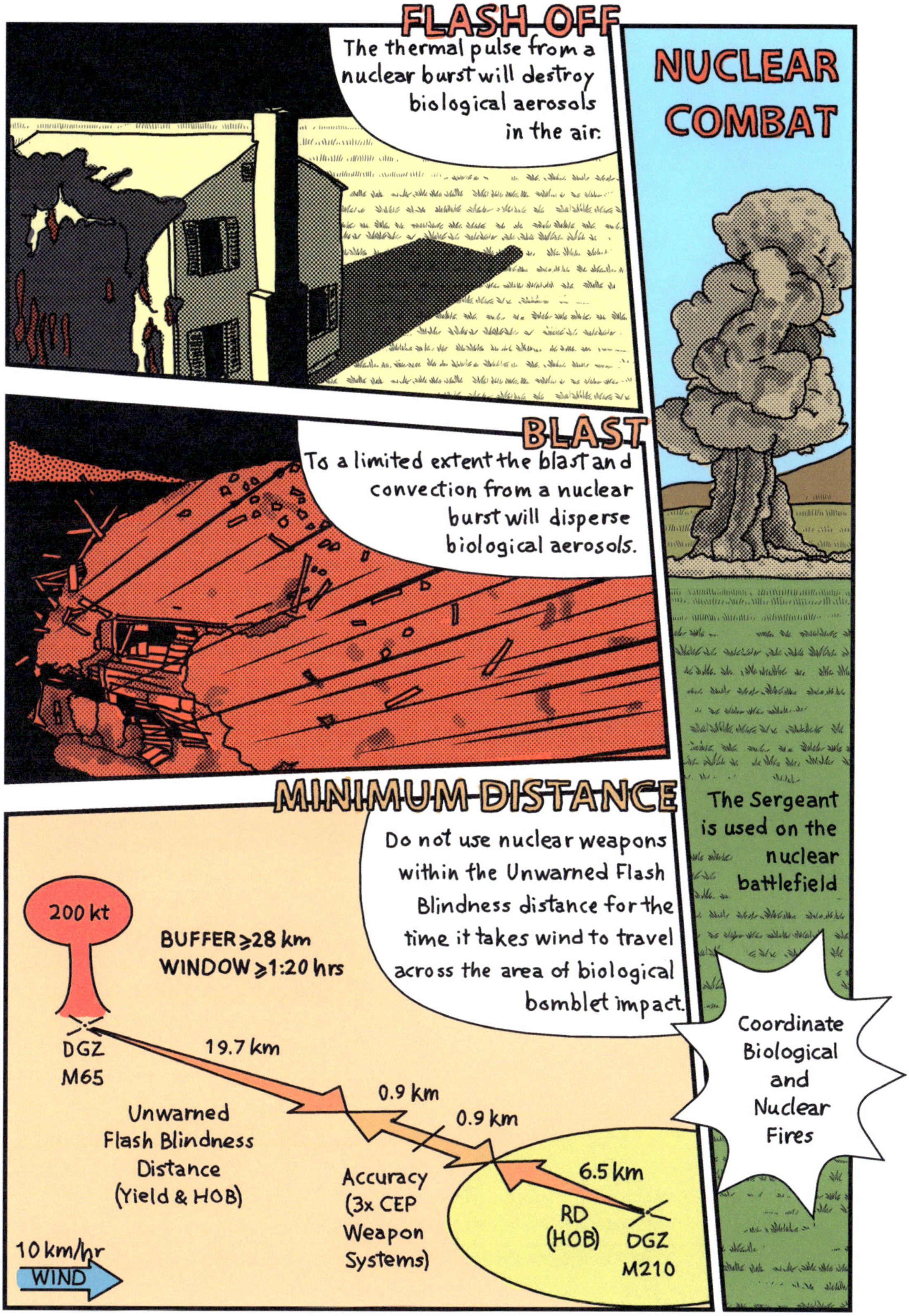
FLASH OFF
The thermal pulse from a nuclear burst will destroy biological aerosols in the air.
NUCLEAR COMBAT
BLAST
To a limited extent the blast and convection from a nuclear burst will disperse biological aerosols.
MINIMUM DISTANCE
Do not use nuclear weapons within the Unwarned Flash Blindness distance for the time it takes wind to travel across the area of biological bomblet impact.
200 kt
BUFFER ≥28 km
WINDOW ≥1:20 hrs
DGZ
M65
19.7 km
Unwarned Flash Blindness Distance (Yield & HOB)
0.9 km
0.9 km
Accuracy (3x CEP Weapon Systems)
6.5 km
RD (HOB)
DGZ
M210
10 km/hr
WIND
The Sergeant is used on the nuclear battlefield
Coordinate Biological and Nuclear Fires

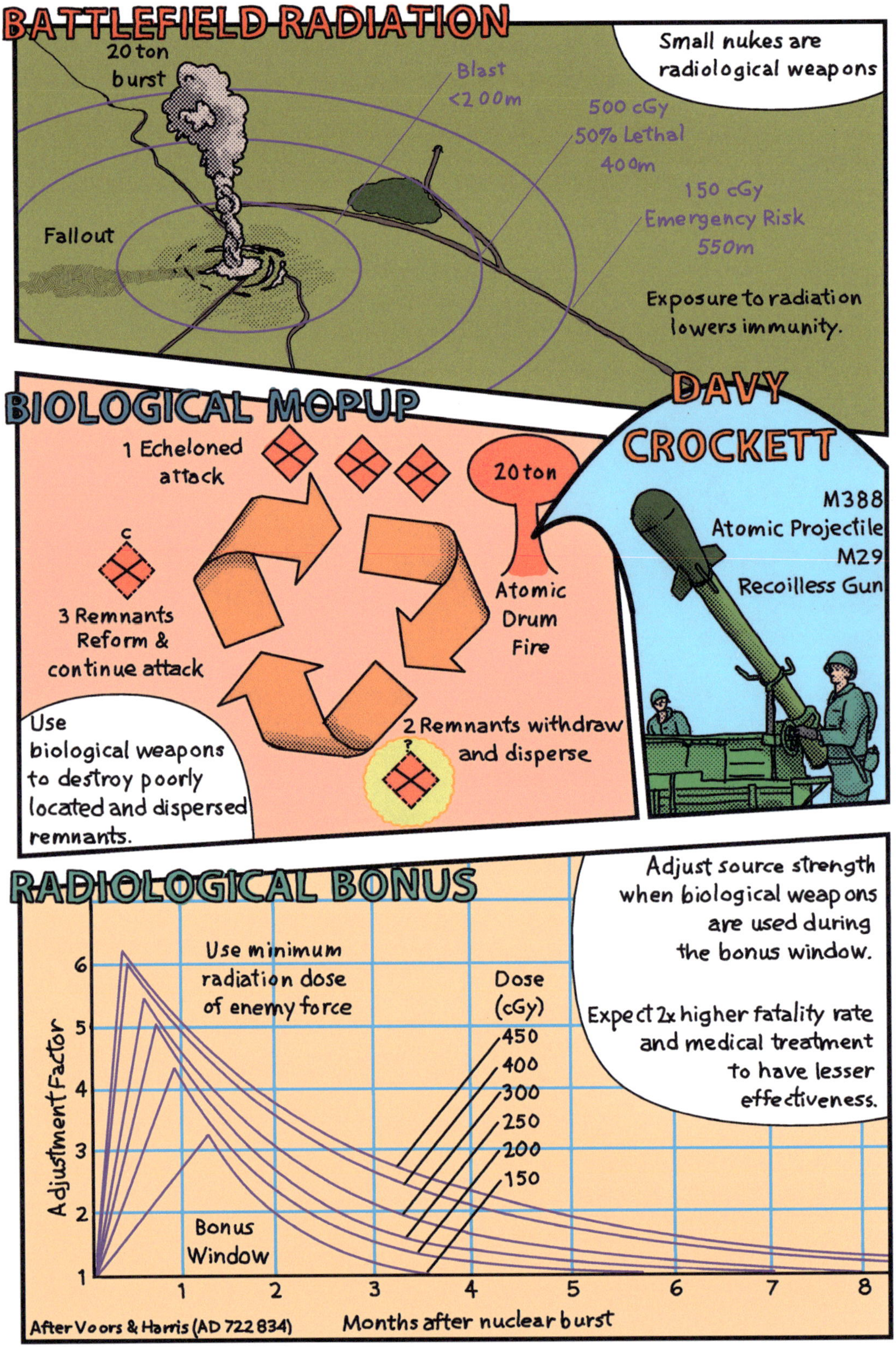
BATTLEFIELD RADIATION
20 ton burst
Blast <200m
Small nukes are radiological weapons
500 cGy 50% Lethal 400m
150 cGy Emergency Risk 550m
Fallout
Exposure to radiation lowers immunity.
BIOLOGICAL MOPUP
1 Echeloned attack
20 ton
Atomic Drum Fire
c
3 Remnants Reform & continue attack
2 Remnants withdraw and disperse
?
Use biological weapons to destroy poorly located and dispersed remnants.
DAVY CROCKETT
M388 Atomic Projectile
M29 Recoilless Gun
RADIOLOGICAL BONUS
Adjust source strength when biological weapons are used during the bonus window.
Expect 2x higher fatality rate and medical treatment to have lesser effectiveness.
Use minimum radiation dose of enemy force
Dose (cGy)
450
400
300
250
200
150
Bonus Window
Adjustment Factor
1
2
3
4
5
6
1
2
3
4
5
6
7
8
Months after nuclear burst
After Voors & Harris (AD 722 834)

# DRY-TYPE AGENTS

The X-1009 facility at the DBO processes frozen pellets of TT through its semi-continuous freeze driers, then mills the product through chilled air-grinders into a 70 µm diameter to a final 5.5 µm diameter product. The final ZZ product is treated with anti-agglomerates and degaussed before packing in dry nitrogen for storage.

# FLETTNER ROTORS

Flettner rotor bomblets have a lift-to-drag ratio of 0.97 and a terminal velocity of 60 to 70 feet per second. They are more self-dispersing than the M143, dispersing up to 46° from vertical.

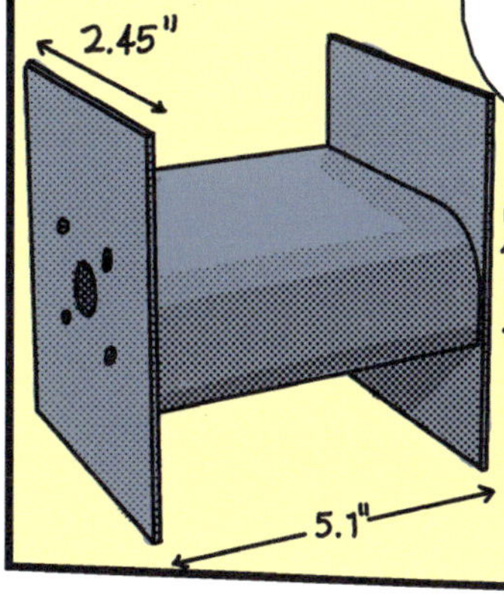


| Dissemination Method | Agent Content (g) | Dissemination Efficiency (%) |
|---|---|---|
| Gas Expulsion | 18 | 30 |
| Telescoping | 48 | 40-60 |

# COVERAGE

A redesign of the M210 may hold 340 or more 5.1-inch Flettner Rotor bomblets. Assuming 95% of these bomblets function, the coverage is—

| HOB (Feet) | RD (Km) | Area ($km^2$) | Density (bomblets/$km^2$) |
|---|---|---|---|
| 25,000 | 7.9 | 196 | 1.7 |
| 30,000 | 9.5 | 282 | 1.1 |
| 35,000 | 11.0 | 383 | 0.8 |
| 40,000 | 12.6 | 501 | 0.6 |
| 45,000 | 14.2 | 634 | 0.5 |
| 50,000 | 15.8 | 782 | 0.4 |

# BOMBLETS OF THE FUTURE

In the next few years biological weapons will cover larger areas with more infective doses.

The BWL is working on an improved M210.

| BIOLOGICAL AGENT | ALTERNATIVES | | FORMULATION Wet-Type | FORMULATION Dry-Type |
|---|---|---|---|---|
| **SR**<br>LETHAL<br>30% Lethality<br>(2 weeks) | **Tularemia** ID50: 100 org/person<br>*Francisella tularensis*<br>Rate-Of-Action: 2-5 Days<br>Duration-of-Action:<br>2 months (Untreated)<br>2 weeks (Treated)<br>Vaccine Available | | **TT** | **ZZ** |
| | | Production Strength | $3.5 \times 10^{10}$ org/ml | $4.5 \times 10^{11}$ org/g |
| | | Storage Half-Life | 40 Days at 41°F | 6 months at 0°F |
| | | Dissemination Efficiency | 0.2% M143 | 30%-60% Flettner |
| **TR**<br>LETHAL<br>>90% Lethality<br>(1 week) | **Anthrax** ID50: 8,000 spores/person<br>*Bacillus anthracis*<br>Rate-of-Action: 1-3 Days<br>Duration-of-Action:<br>1-3 weeks<br>(Treatment Ineffective)<br>Vaccine Available | | **TR1** | **TR2** |
| | | Production Strength | $3.5 \times 10^{9}$ spores/ml | $3.0 \times 10^{11}$ spores/g |
| | | Storage Half-Life | 18 months at 70°F | 18 months at 70°F |
| | | Dissemination Efficiency | 4% M143 | 30%-60% Flettner |
| **LM**<br>INCAPACITANT<br>1% Lethality | **Q Fever** ID50: 18 GPU/person<br>*Coxiella brunettii*<br>Rate-of-Action: ~2 weeks<br>Duration-of-Action:<br>4 weeks (Untreated)<br>1 Week (Treated) | | **MN** | **NT** |
| | | Production Strength | $1.0 \times 10^{10}$ GPU/ml | $1.0 \times 10^{12}$ GPU/g |
| | | Storage Half-Life | 5 months at 41°F | 30 months at 0°F |
| | | Dissemination Efficiency | 0.3% M143 | 30%-60% Flettner |
| **DK**<br>INCAPACITANT<br>0% Lethality | **Venezuelan Equine Encephalomyelitis** ID50: 10 MU/person<br>Rate-of-Action: 2-5 Days<br>Duration-of-Action:<br>1-2 weeks<br>(No Treament Available)<br>Vaccine Available | | **FX** | **TD** |
| | | Production Strength | $2.0 \times 10^{10}$ MU/ml | $3.0 \times 10^{11}$ MU/g |
| | | Storage Half-Life | 7 months at -40°F | 6 months at 41°F |
| | | Dissemination Efficiency | 2% M143 | 30%-60% Flettner |
| **OZ**<br>INCAPACITANT<br>2% Lethality | **Porcine Brucellosis**<br>*Brucella suis* ID50: 500 org/person<br>Rate-of-Action: ~2 weeks<br>Duration-of-Action:<br>3 months (Untreated)<br>1 month (Treated) | | **NX** | |
| | | Production Strength | $3.0 \times 10^{10}$ org/ml | |
| | | Storage Half-Life | 2 months at 41°F | |
| | | Dissemination Efficiency | 0.7% M143 | |

DBO can change production line within 1 month

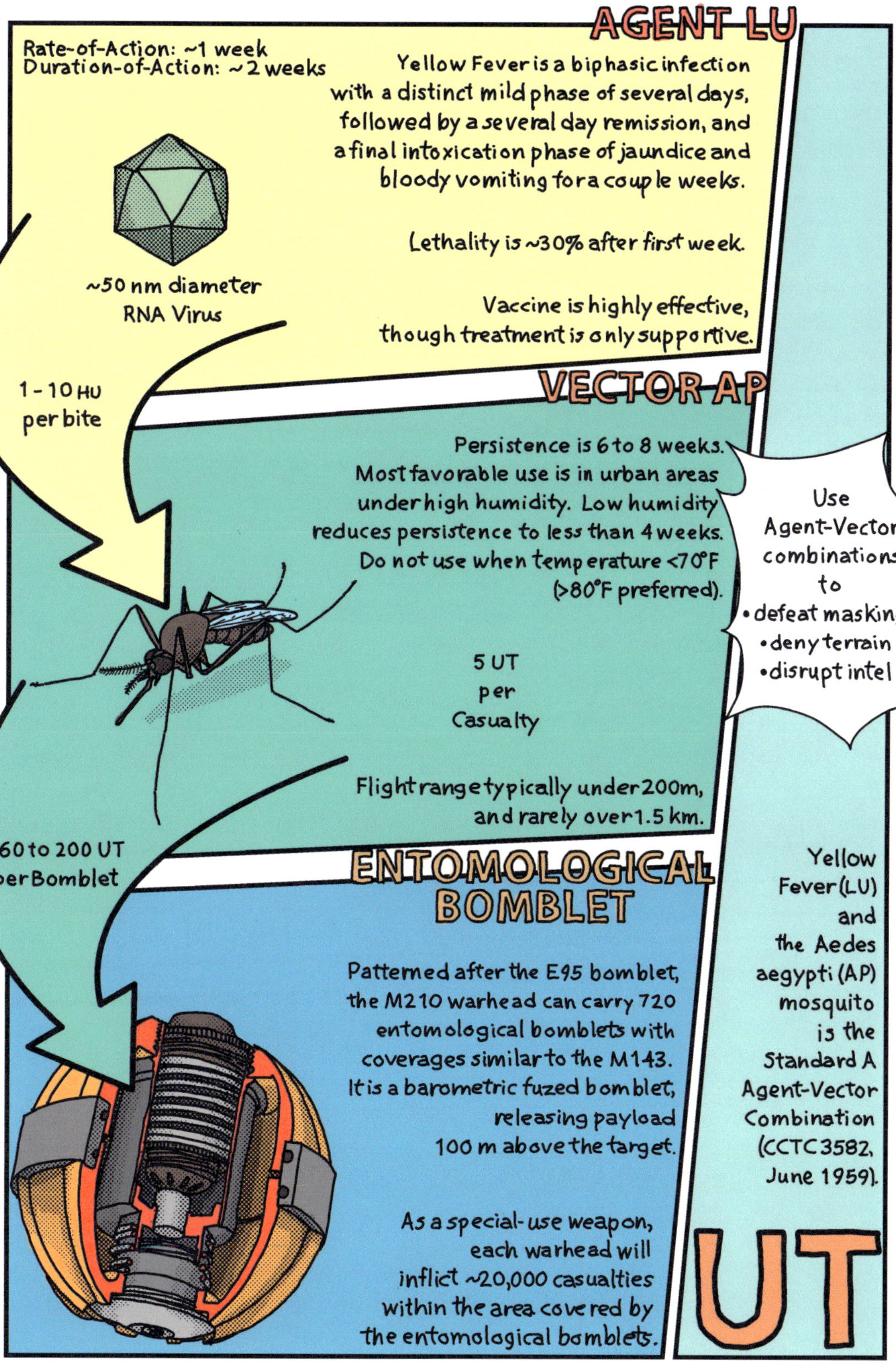
AGENT LU
Rate-of-Action: ~1 week
Duration-of-Action: ~2 weeks
Yellow Fever is a biphasic infection with a distinct mild phase of several days, followed by a several day remission, and a final intoxication phase of jaundice and bloody vomiting for a couple weeks.
Lethality is ~30% after first week.
Vaccine is highly effective, though treatment is only supportive.
~50 nm diameter RNA Virus
1 - 10 HU per bite
VECTOR AP
Persistence is 6 to 8 weeks. Most favorable use is in urban areas under high humidity. Low humidity reduces persistence to less than 4 weeks. Do not use when temperature <70°F (>80°F preferred).
5 UT per Casualty
Flight range typically under 200m, and rarely over 1.5 km.
Use Agent-Vector combinations to
• defeat masking
• deny terrain
• disrupt intel
160 to 200 UT per Bomblet
ENTOMOLOGICAL BOMBLET
Patterned after the E95 bomblet, the M210 warhead can carry 720 entomological bomblets with coverages similar to the M143. It is a barometric fuzed bomblet, releasing payload 100 m above the target.
As a special-use weapon, each warhead will inflict ~20,000 casualties within the area covered by the entomological bomblets.
Yellow Fever (LU) and the Aedes aegypti (AP) mosquito is the Standard A Agent-Vector Combination (CCTC 3582, June 1959).
UT

# LOGISTICS

UT is produced by the DBO's X-1001 facility and supplied in lots of 100 per a gallon carton. UT has a half-life of 40 days when maintained at 80°F, 80% humidity, and 10% sucrose feedings every 48 hours.

(1 warhead requires ~150,000 live UT)

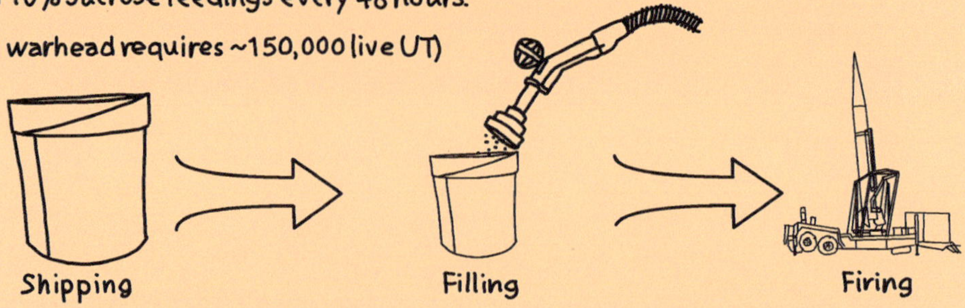


Entomological bomblets are filled by vacuuming live UT from cartons in-Theater. A UT loaded warhead must be used within 24 hours after filling.

# TROOP SAFETY

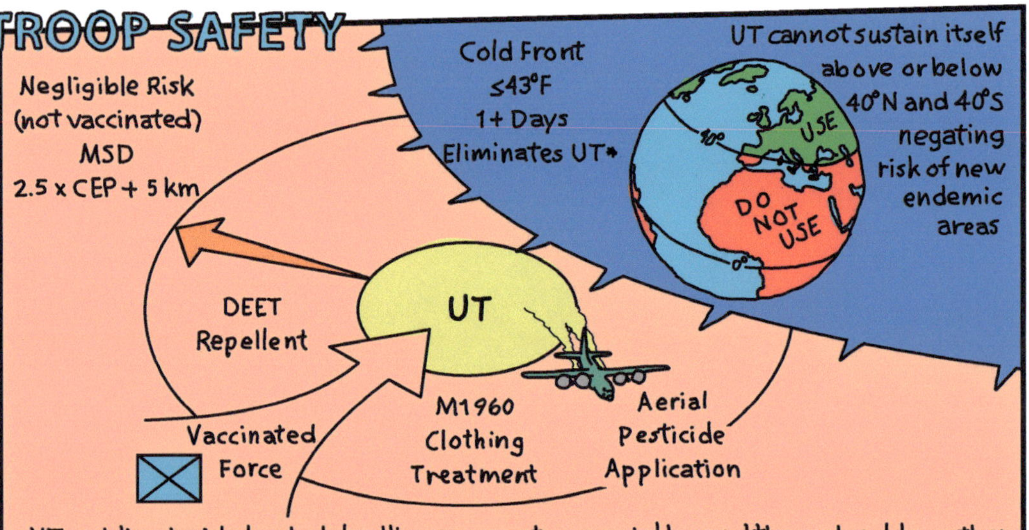


*UT residing inside heated dwellings presents a special hazard through cold weather.

# FUTURE COMBINATIONS

In the future, other Agent-Vector combinations using AP may become available. These alternatives to UT have a similar envelope-of-action and persistence. They differ in lethality and availability of vaccines.

| Agent | Biological | Lethality (%) |
|---|---|---|
| RR | Chikungunya | 0 |
| ZQ | O'nyong-nyon | 0 |
| FA | Rift Valley Fever | 4 |
| EV | Western Equine Encephalitis | 4 |
| ZX | Eastern Equine Encephalitis | 33 |
| AN | Japanese Encephalitis | 50 |

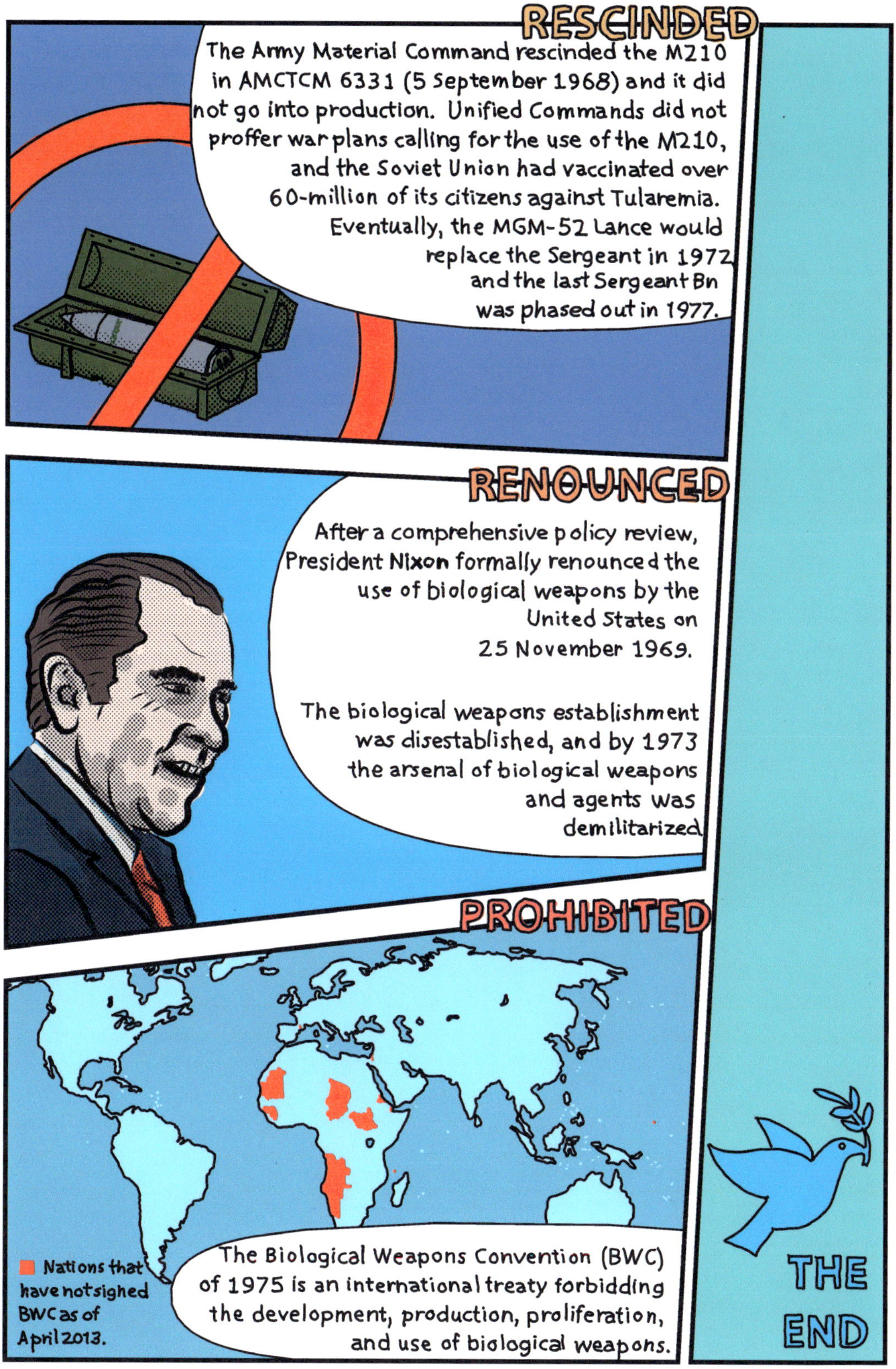
RESCINDED
The Army Material Command rescinded the M210 in AMCTCM 6331 (5 September 1968) and it did not go into production. Unified Commands did not proffer war plans calling for the use of the M210, and the Soviet Union had vaccinated over 60-million of its citizens against Tularemia. Eventually, the MGM-52 Lance would replace the Sergeant in 1972 and the last Sergeant Bn was phased out in 1977.
RENOUNCED
After a comprehensive policy review, President Nixon formally renounced the use of biological weapons by the United States on 25 November 1969.
The biological weapons establishment was disestablished, and by 1973 the arsenal of biological weapons and agents was demilitarized.
PROHIBITED
Nations that have not sighed BWC as of April 2013.
The Biological Weapons Convention (BWC) of 1975 is an international treaty forbidding the development, production, proliferation, and use of biological weapons.
THE END

Made in the USA
Monee, IL
05 November 2026

67953743R00043